Materials Science: A Field of Diverse Industrial Applications

Edited by

Arti Srivastava

Department of Chemistry, Guru Ghasidas Vishwavidyalaya
Central University, Bilaspur, CG 495009, India

Mridula Tripathi

Department of Chemistry, C.M.P. Degree College, University
of Allahabad, Uttar Pradesh 211002, India

Kalpana Awasthi

Department of Physics, K. N. Govt. P.G. College, Purvanchal
University, Uttar Pradesh 222003, India

&

Subhash Banerjee

Department of Chemistry, Guru Ghasidas Vishwavidyalaya,
Central University , Bilaspur-495009, CG, India

Materials Science: A Field of Diverse Industrial Applications

Editors: Arti Srivastava, Mridula Tripathi, Kalpana Awasthi and Subhash Banerjee

ISBN (Online): 978-981-5051-24-7

ISBN (Print): 978-981-5051-25-4

ISBN (Paperback): 978-981-5051-26-1

First published in 2023.

need for a court order if at any point you breach any terms of this License Agreement. In no event will any delay or failure by Bentham Science Publishers in enforcing your compliance with this License Agreement constitute a waiver of any of its rights.

3. You acknowledge that you have read this License Agreement, and agree to be bound by its terms and conditions. To the extent that any other terms and conditions presented on any website of Bentham Science Publishers conflict with, or are inconsistent with, the terms and conditions set out in this License Agreement, you acknowledge that the terms and conditions set out in this License Agreement shall prevail.

Bentham Science Publishers Pte. Ltd.
80 Robinson Road #02-00
Singapore 068898
Singapore
Email: subscriptions@benthamscience.net

BENTHAM SCIENCE

CONTENTS

PREFACE

Material science comprises various classes of different materials and every class constitutes a separate field. New materials are developed through nanotechnology in various fields of material science that develop areas of research like biotechnology, environmental science, information technology and energy technologies. The present book attempts to cover recent developments of new materials and their application in various fields. This book consists of ten book chapters from well-reputed universities and institutes. Chapters of the book cover areas like nanomaterials in society, nanomaterials in medicine in health sectors, and nanomaterials as nano food. Some chapters of this book cover the application of materials/ nanomaterials for energy and their functionalization for particular applications like multifunctional advanced nanomaterials for energy applications, materials in dye-sensitized solar cells and for electrochemical sensing. The content of the book has three sections. A chapter-by-chapter brief description is as follows:

The first sections have five chapters related to the different modified materials and their versatility in different fields of science. In this respect, the authors of chapter 1 have discussed the silent characteristics of many inorganic materials and their medicinal significance. Chapter 2 highlights some basics of chalcogenide glasses, preparation techniques and a review of the latest technological developments along with structural properties, optical properties, and thermal and electrical properties of chalcogenide glasses. The author presents the method of fabrication of chalcogenides, in bulk and thin film forms in Chapter 3. Dr. Sunanda Das has provided a detailed study in chapter 4 about polymeric synthetic fibers and their commercial uses. She has also highlighted the points related to the environmental impact and health issues engendered by using a lot of synthetic polymeric fibres in different areas. Chapter 5 presents detailed surface morphological studies using AFM images of BaF_2 thin films deposited by the electron beam evaporation technique on three different substrates like glass, silicon (Si), and aluminum (Al) substrate and also estimates the fractal dimensions of the horizontal as well as vertical sections of the film surfaces.

The second section of the book content has three chapters on functionalized nanomaterials and also their applications. Chapter 6 deals with the synthesis and characterization of BFO and Sm-doped BFO nanoceramics. Chapter 7 presents a review of the development of nano-technology for the removal and safe disposal of radioactive ions from the environment using nanomaterials. Chapter 8 is the ephemeral study of nanocatalysts in organic synthesis.

The third section of the book contains chapters related to the role of modified materials in energy production and its utilization. In Chapter 9, *Pinki, Subhash* and *Ashu Chaudhary* have given an overview of the implications and applications of multifunctional advanced materials/gadgets for energy conversion and storage. They have explained different processes used for energy conversion and storage like lithium-ion batteries, supercapacitors, fuel cells, polyoxometalates, polyaniline-based two-dimensional dichalcogenides, chemical vapour deposition, thermal energy storage with phase change materials, *etc*. Chapter 10 describes the overall advancement in electrolytes used for producing low-cost and industrially stable dye-sensitized solar cells. For this purpose, Priyanka Chawla, *et al*. have explained the use of solid polymeric materials such as PEO, PVA, PVDF, and chitosan in place of liquid electrolytes.

Arti Srivastava
Department of Chemistry
Guru Ghasidas Vishwavidyalaya Central University
Bilaspur, CG 495009
India

Mridula Tripathi
Department of Chemistry
C.M.P. Degree College
University of Allahabad
Uttar Pradesh 211002, India

Kalpana Awasthi
Department of Physics
K. N. Govt. P.G. College
Purvanchal University
Uttar Pradesh 211002, India

&

Subhash Banerjee
Department of Chemistry
Guru Ghasidas Vishwavidyalaya
Central University , Bilaspur-495009, CG
India

List of Contributors

Anchal Srivastava	Department of Physics, University of Lucknow, Lucknow, India
Achchhe Lal Saroj	Department of Physics, Institute of Science, Banaras Hindu University, Varanasi, UP, India
Anil Kumar	Department of Physics & Electronics, Dr. Ram Manohar Lohia Avadh University, Ayodhya, India
Amreesh Chandra	Department of Physics, Indian Institute of Technology Kharagpur, Kharagpur, India
Akram Ali	Department of Chemistry, CMP Degree College, University of Allahabad, Uttar Pradesh 211002, India
Ashu Chaudhary	Department of Chemistry, Kurukshetra University, Kurukshetra, Haryana, India
Deepanjali Pandey	Department of Chemistry, CMP Degree College, University of Allahabad, Uttar Pradesh 211002, PrayagrajIndia
Dharamveer Singh	School of Sciences, Rajshree Tondon Open University, Prayagraj, India
Horesh Kumar	Department of Physics, Institute of Science, Banaras Hindu University, Varanasi, UP, India
Hari Pratap Bhasker	Department of Physics, Chaudhary Mahadeo Prasad Degree College, University of Allahabad, Prayagraj, India
Kamakhya Prakash Misra	Department of Physics, School of Basic Sciences, Manipal University Jaipur, Jaipur, Rajasthan, India
Kumari Pooja	Department of Chemistry, CMP Degree College, University of Allahabad, Uttar Pradesh 211002, India
Monika Singh	Department of Chemistry, CMP Degree College, University of Allahabad, Uttar Pradesh 211002, India
Manoj Kumar	Department of Chemistry, CMP Degree College, University of Allahabad, Uttar Pradesh 211002, India
Nishant Kuma	Amiruddaula Islamia Degree College, Lucknow, India
Priyanka Srivastava	Institute of Engineering and Technology, Dr. Ram Manohar Lohia Avadh University, Ayodhya, India
Pramesh Chandra	Indira Gandhi National Open University, Khanna, India
Pradip Kumar Priya	Department of Physics, Ewing Christian College, University of Allahabad, Prayagraj, India
Kusum Lata Pandey	Department of Physics, Ewing Christian College, University of Allahabad, Prayagraj, India
Pinki Subhash	Department of Chemistry, Kurukshetra University, Kurukshetra, Haryana, India
Priyanka Chawla	Department of Chemistry, CMP Degree College, University of Allahabad, Uttar Pradesh 211002, India

Ram Pratap Yadav   Department of Physics, Deen Dayal Upadhyay Govt. P.G. College, Saidabad, Prayagraj, India

Raj Kumar Anand   Department of Physics, University of Allahabad, Prayagraj, India

Rajesh Kumar Shukla   Department of Physics, University of Lucknow, Lucknow, India

Satish Kumar Mandal   Department of Physics, Indian Institute of Technology Kharagpur, Kharagpur, India
Surface Physics and Material Science Division, Saha Institute of Nuclear Physics, Kolkata, India

Savita   Department of Physics, University of Allahabad, Prayagraj, India
Government Girls Polytechnic Ayodhya, Uttar Pradesh, India

Shalini Verma   Department of Chemistry, University of Allahabad, Prayagraj, India

Shraddha Tivari   Department of Chemistry, CMP Degree College, University of Allahabad, Uttar Pradesh 211002, India

Seraj Ahmad   Department of Chemistry, CMP Degree College, University of Allahabad, Uttar Pradesh 211002, India

Shivangi Trivedi   Department of Chemistry, CMP Degree College, University of Allahabad, Uttar Pradesh 211002, India

Sunanda Das   Department of Chemistry, Chaudhary Mahadeo Prasad Degree College, University of Allahabad, Prayagraj, Uttar Pradesh, India

Vishal Srivastava   Department of Chemistry, CMP Degree College, University of Allahabad, Uttar Pradesh 211002, India

Vijay Krishna   Department of Chemistry, University of Allahabad, Prayagraj, India

Part 1: Functional Materials and their Versatility

CHAPTER 1

Chalcogenides: Bulk and Thin Films

Anchal Srivastava[1,*], Rajesh Kumar Shukla[1], Priyanka Srivastava[2], Pramesh Chandra[3] and **Nishant Kumar[4]**

[1] *Department of Physics, University of Lucknow, Lucknow, India*

[2] *Institute of Engineering and Technology, Dr. Ram Manohar Lohia Avadh University, Ayodhya, India*

[3] *Indira Gandhi National Open University, Khanna, India*

[4] *Amiruddaula Islamia Degree College, Lucknow, India*

Abstract: Chalcogens are the chemical elements of group 16 of the periodic table. Oxygen is treated separately from other chalcogens; it is even excluded from the term 'chalcogen' altogetherdue to its very different chemical behaviour from sulfur, selenium, tellurium and polonium. The heavier chalcogens have vacant d orbitals. A chalcogenide consists of at least one chalcogen element and one electropositive element. The term chalcogenide is more commonly reserved for sulfides, selenides and tellurides rather than oxides. The interest in these materials arises particularly due to their ease of fabrication in the form of bulk and thin films. Generally, chalcogenides have a large glass-forming region and thus, their physical properties can be tuned *via* chemical composition. These glasses have drawn great attention due to their striking electrical, optical and thermal properties, which ary with composition, heat treatment, irradiation, glass forming methods, *etc*. There is a high tendency for the atoms to link together to form link chains in chalcogenides. In general, the atomic bonding is more rigid than that of organic polymers and more flexible than that of oxide glasses. This chapter presents the method of fabrication of chalcogenides in bulk and thin film forms.

Keywords: Bulk form, Chemical methods, Chalcogenides, Physical deposition, Synthesis, Thin films.

CHALCOGENS

Chalcogens are the chemical elements of group 16 of the periodic table. The group consists of the elements oxygen (O), sulfur (S), selenium (Se), tellurium (Te) and polonium (Po) each having six valence electrons in the outermost shell.

---

[*] **Corresponding author Anchal Srivastava:** Department of Physics, University of Lucknow, Lucknow, India; E-mail: asrivastava.lu@gmail.com

Arti Srivastava, Mridula Tripathi, Kalpana Awasthi and Subhash Banerjee (Eds.)

The term "chalcogens" was derived from the Greek word *chalcos* meaning 'copper or ore' and *genes* meaning 'formed or produced' since they all are found in copper ores. The name became popular since it is analogous to the name of group 17 – halogens – meaning salt formers. Oxygen is a gas while the other group members are solids. Both oxygen and sulfur can be found in the pure form. Oxygen is generally extracted from air and sulfur is extracted from oil and natural gas. Selenium and tellurium are produced as byproducts of copper refining. Polonium – a radioactive element – is mostly available in particle accelerators. Oxygen is treated separately from other chalcogens; even excluded from the scope of the term "chalcogen" altogether, due to its very different chemical behaviour from sulfur, selenium, tellurium and polonium. One reason for this is that the heavier chalcogens have vacant d-orbitals. Oxygen's electro-negativity is also much higher than those of other chalcogens. This makes oxygen's electric polarizability several times lower than those of other chalcogens [1]. The metal character increases and the electro-negativity decreases, as the atomic number increases. The elements and their compounds vary in their toxicity. It is interesting that oxygen and sulfur are essential to all life, while the compounds of selenium, tellurium and polonium can be toxic. The chalcogens have varying crystal structures. Oxygen's crystal structure is monoclinic while sulfur's is orthorhombic. Selenium and tellurium have a hexagonal crystal structure while polonium has a cubic crystal structure.

CHALCOGENIDES

A chalcogenide is an inorganic chemical compound which consists of at least one chalcogen element and at least one more electropositive element. Although all group 16 elements of the periodic table are termed as chalcogens, chalcogenide is more commonly used for sulphides, selenides and tellurides only, rather than oxides. Oxides behave quite differently from chalcogenides. Their band gaps and optical and electrical properties are very different. Hence, chalcogenides may be defined as oxygen-free inorganic materials which contain one or more chalcogen elements along with at least one electropositive element. Chalcogenides possess large glass-forming region so their properties – physical, electrical, optical and thermal – can be tuned by altering the chemical composition, irradiation, heat treatment, glass forming method, *etc.* It is the chemical bonding in combination with the topology of glass network that determines their physical properties. Therefore, it is necessary that structure of these glasses is properly studied so as to form compositions with the best performance. Chalcogenides draw attention due to their properties which are found to vary with composition, Atoms have high tendency to link together to form link-chains in chalcogenides. Short-range inter-atomic forces are mainly covalent. These are highly directional and strong in magnitude. The weak van der waals forces contribute primarily to the medium-range order. By and large, the atomic bonding is comparatively more rigid than that of organic polymers whereas it is found to be more flexible than that of the oxide glasses. Due to the weak van der waals bonding between layers/chains, there is flexibility of structure which means changes in the structure can be done easily [2].

Since each atom needs two neighbours to fulfil the valence requirements, it can adjust its neighbouring environment. Amorphous selenium in its pure form consists of a mixture of two structural species – long helical chains and eight member rings (Se_8) – held to each other by weak van der waals forces. Additives such as Te, Sb, Cu, Pb, Cd, Bi, In, Ge, *etc.* are used with Se to form binary as well as ternary chalcogenide alloys for enhanced performance. Se-Te alloys are more advantageous than amorphous selenium (a-Se) due to their greater hardness, higher photosensitivity, higher crystallization temperature and smaller ageing effects [3]. The structure of a-Se consists of two-fold co-ordinations and has long polymeric Se_n chains. Addition of tellurium (Te) produces a catalytic effect on the crystallization of selenium. The presence of tellurium in Se chains favours their thermal dissociation, since the Se-Te bond is weaker than Se-Se bond. In Se-Te alloys, Te enters the structure by isoelectronic substitution and consequently changes the inter-chain van der waals bonding since the atomic size of Te is larger than that of Se. The addition of Te in Se also improves the corrosion resistance and reduces the band gap of Se [4]. Se-Te alloys can be used for fast change between amorphous and crystalline phases. But, these alloys have some significant setbacks during certain applications such as limited reversibility, low glass-transition (T_g) and crystallization (T_c) temperatures, *etc.* These difficulties can be removed by adding a third element in Se–Te binary alloys. An additive in Se-Te alloys behaves as a modifier and changes various properties *viz.* optical, thermal, electrical *etc.* of the host alloy. The cross-linkages between the chains strengthen the material. It can also create structural and configurational changes in the system. The conductivity type may switch from p to n [5]. A typical chalcogenide exhibits sharp optical absorption edge, single electrical activation energy, efficient photo-excited conductivity and luminescence. It has been found that elements, such as Cd, Sb, Sn, In, Pb, Bi, Ag, *etc.* when added in Se-Te alloy produce noticeable changes in the structural, optical, electronic and thermal properties [6 - 12]. However, reports about copper as metal additive are scarce in literature. Various examples of chalcogenides occurring in nature are PbS, PbSe, PbTe, HgS, ZnS, CdS, CdSe, Ag_2S, Ag_2Te, Cu_2S, Cu_2Se, As_2S_3, As_4S_4, Sb_2S_3, Bi_2S_3, Bi_2Te_2S, $Pb_2Sb_2S_4$, $Pb_5Sb_4S_{11}$, Cu_3AsS_4, GeTe, MoS^2, *etc.* and those synthesised are crystals or thin films of CaS, PbS, PbSe, PbTe, HgTe, CdTe, Se, As_2Se_3, As_2Se_3, Ge-Sb-Se, Ge-Sb-S and Ge-La-S systems, Sb_2S_3, Sb_2Te_3, Sb_2Se_3, Bi_2S_3, Bi_2Se_3, Bi_2Te_3, Ge-Sb-Se, Ge-Bi-S, Ge-Bi-Se, $Ge_2Sb_2Te_5$, $Ge_8Sb_2Te_{11}$, $Ge_8Bi_2Te_{11}$, *etc.*

APPLICATIONS OF CHALCOGENIDES

Chalcogenides are a very important class of materials because of their optical, electrical and technological applications. These materials are useful for applications in the infrared region due to their wide transmission range, good chemical durability, high refractive index and flexibility. Selenide and telluride alloys in binary or multi-component systems are utility materials in the spectral range 600 – 1500 nm for various optical and photonic applications. These are also important for applications as passive devices such as lenses, windows, fibres, *etc.* as well as for preparation of active devices such as laser fibre amplifiers, non-linear components, *etc.*

Chalcogenides are low-phonon energy materials and are generally transparent from the visible to infrared region [13]. Waveguides and fiber structures can be developed because of their sensitivity to absorption of electromagnetic radiation with a variety of photoinduced effects [14]. Optical recording of information has got great importance. Selenium is quite useful material for the development of solid-state devices, such as switching and memory devices. It is used as photoconductors in high-definition television (HDTV) [15] and digital radiography (DDR) [16]. This is because of the production of low thermal noise, high spatial resolution and high sensitivity in the wide range of wavelengths from visible to ultraviolet as well as x-rays as compared to silicon based photoconductors. Selenium alloys also show property of reversible transformation which makes these systems quite useful for optical memory, X-ray imaging and photonics [17]. Selenides have potential applications in thin film solar cells [18], photoelectrochemical cells [19], heterojunction solar cells [20], photodetectors [21], photoconductive cell, *etc.* These are also used as high performance counter electrodes in dye sensitized solar cells [22], catalysts, waste water treatments, optical coatings [23] and photoelectrodes [24].

Se-Te binary alloys have been widely studied due to their (a) great storage capacity and fast-to-access information, (b) advantage to delete and introduce new information and (c) electrophotographic applications such as photocopying and laser printing, phase change recording *etc* [25]. Addition of tellurium in selenium helps in reducing the band gap of Se so as to make it useful in photoreceptors [4]. However the band gap may be differently affected in case of nanoparticles. On the other hand, tellurium-rich alloys are also attracting a lot of attention as they are being considered for their potential applications in data storage devices [26]. It has been found that tellurium-rich alloys also have good transparency in the infrared range and have high refractive index, which render these systems suitable for optical devices. Applications of tellurium-rich Se-Te alloy range from optical recording media to xerography. These materials are sensitive to laser irradiation. The absorption of laser irradiation in chalcogenides depends on their electronic structure, which in turn changes by interaction with photons. Other applications include phase-change materials [27], sensors, optical circuits, gratings, waveguides [28], *etc.*

Chalcogenides are remarkably useful in infrared optics. These are well known for their transparency far into the infrared and are widely used as infrared optical components which have commercial utility. The development of infrared (IR) technologies such as night vision cameras and bio-optical sensors in recent decades has gained increasing interest so far as investigations of materials used for IR fibers with low attenuation are concerned. Ideal candidate materials for this should have specified features such as an exceedingly wide transparency window, good thermal properties for fiber drawing, good chemical stability *etc.* Chalcogenides have been found useful for such applications because of their wide transparent window that covers the atmospheric IR windows of 3000 – 5000 nm and 8000 – 12000 nm. Stronger glass forming tendency, better stability against crystallization and chemical durability to resist corrosion make them superior as compared to other non-oxide materials [29]. Chalcogenide fibers are well-suited for chemical sensor applications such as fiber-optic chemical sensor systems for quantitative remote detection and identification as well as for detection of chemicals in mixtures. These are also used in fabricating many devices such as waveguides, fiber Bragg gratings, nonlinear directional couplers, *etc.* using light-induced photo-structural changes [2]. Applications of IR optics also include energy management, remote sensing, thermal fault detection, temperature monitoring, electronic circuit detection, infrared laser power transmission and scanning near-field IR microscopy. There are analytical applications of chalcogenide glasses as chemical sensors in environmental monitoring and process control and as a semiconductor-based field-effect platform for bio-chemical and physical sensors.

The phenomenon related to electronic conduction in amorphous chalcogenide semiconductors has attracted a great deal of scientific attention since the discovery of electrical switching in chalcogenides glasses in 1968. Glasses exhibiting no crystallization reaction above T_g show threshold switching, whereas those having crystallization reaction above T_g exhibit memory switching. Memory switches are formed at the boundaries of glass-forming regions where glasses are stable and have the tendency to crystallize when heated or cooled slowly [30].

The property of optical non-linearity has also been gaining increasing recognition and has led to a number of demonstrations of all-optical processes including switching [31], regeneration, wavelength conversion [32], amplification, lasing, pulse compression and slow light [33].

Another important property that the chalcogenides show is photosensitivity to band-gap light. Apart from bulk optic components, chalcogenide glass fiber – step index as well as microstructured – and planar waveguide devices have been developed. Optoelectronic applications include photo-detectors, gas sensors, blue and white light emitting diodes (LEDs), thin film transistors, optical wave guides and solar cells, blue diode lasers, *etc.* Numerous other applications are X-ray sensors, Xerox facilities, ferroelectrics, thermoelectric, catalysers, plasmonic materials, multiphoton process, nanodots, optomechanical devices, surgical instruments, *etc.*

MATERIALS AND METHODS

A chalcogenide material may be expressed by the formula MX where M is any metal and X is a chalcogen. Also, M can be a combination of metals (Ti, V, Cr, Mn, Cu, Zn, Cd, Pb, *etc.*) and X can be a combination of chalcogens (S, Se, Te). Chalcogenides are substances which have an incongruent melting point and exhibit a high partial vapour pressure during melting. Therefore, synthesis must be carried out keeping in view the conditions that are widely varied and are dependent on the glass composition, glass forming region and glass forming ability. Various forms of chalcogenides include crystalline, amorphous, crystals, thin films, fibres, nanoparticles, *etc.* Out of the vast area of interest, only a few materials and processes can be discussed here.

SYNTHESIS OF BULK CHALCOGENIDES

Over the past few decades, synthesis of chalcogenides has attracted significant interest and is still the subject of investigations. Conventionally, metal chalcogenides can be prepared in a variety of ways – physical, chemical, biological or hybrid. The most common is the direct combination of the elements at elevated temperatures. Irrespective of the method of synthesis of chalcogenides, it is quite important to avoid the coalescence of particles into larger particles and achieve chemical stability.

Melt Quenching Technique

Bulk samples can be synthesized using the well-known and widely used melt quenching technique. This technique makes use of rapid quenching of a melt and is historically the most established and still the most widely used method in the preparation of chalcogenide materials [34]. The method consists of weighing and mixing the high purity constituents followed by melting at high temperature in sealed evacuated silica/quartz ampoules. During the process of heating, the ampoule is frequently rocked so as to ensure homogenization of the melt. It is quenched rapidly by dropping the ampoule immediately from the furnace into a liquid. This liquid should preferably be the one with high thermal conductivity as well as high latent heat of vapourization so that the heat is conducted away from the sample as fast as possible without forming the thermally insulating vapour layer around the ampoule. An ice-water bath is generally used for rapid quenching.

The basic point to consider is to avoid the process of crystallization during quenching. The important factor which affects glass forming ability is the viscosity of the melt. The more viscous the melt at a given temperature above the melting point, the greater is the tendency to form glass. The rate of quenching is another factor that has an effect on the glass forming ability of chalcogenide materials. In general, the faster the rate of quenching of the melt, the greater is the likelihood of forming a glass and not a crystalline product.

Thermolysis or Thermal Decomposition

One of the simplest methods to prepare bulk samples is the decomposition of organometallic precursors. The molecular precursor methods involve thermal decomposition of a compound that contains an M-S or M-Se bond. An advantage of using organometallic compounds is that the precursors can be decomposed at relatively low temperatures (200 – 350°C) to form the final product. By controlling the decomposition temperature, the growth of nanoparticles can be controlled. Studies have been done on nanoparticles of CdSe/CdS, CdSe/ZnS, CdSe/ZnSe core/shell and CdSe/CdS alloys [35]. The advantage of this method is that the reaction does not need any additional reducing agents. This method is cost-effective and also gives high yield with fine grains. The biggest disadvantage of this approach is that most of the reactions involve air-sensitive reactants as well as the final precursor. Therefore, glove box or schlenck line techniques have to be used.

Sonochemical Synthesis or Ultrasound Irradiation

Ultrasound irradiation provides an unusual mechanism to generate high-energy chemistry with extremely high local temperatures and pressures and an extraordinary heating and cooling rate. Sonochemistry drives principally from acoustic cavitations: the formation, growth and implosive collapse of bubbles in liquids. When solutions get exposed to strong ultrasound irradiation, bubbles collapse implosively by acoustic fields in the solution. High temperature and high pressure fields are produced at the centers of the bubbles. This generates a localized hotspot through adiabatic compression or shock wave formation within the gas phase of the collapsing bubbles. Ultrasound irradiation differs from traditional energy sources such as heat, light or ionizing radiation in terms of duration, pressure and energy per molecule. It offers a very attractive method for the preparation of various nano-sized metal chalcogenides. They do vary in size, shape, structure and in their solid phase – amorphous or crystalline – are always of nanometer size.

Photochemical Synthesis

Absorption of photoenergy can change the structure of molecules and induce a variety of photochemical reactions. Photochemical synthesis has also emerged as an effective synthetic procedure for the preparation of nanosized metal chalcogenides with various morphologies. The equipments involved are simple and economic. A low-pressure mercury pillar lamp as the ultraviolet irradiation source and a high-pressure column-like indium lamp as visible photo-irradiation source are most commonly used. This method has the advantages of mild reaction conditions and convenient operations. CdSe nanoparticles have been studied using this method.

Chemical Precipitation/Co-Precipitation

A chemical precipitation process consists of three main steps: chemical reaction, nucleation and crystal growth. By and large, it is not a controlled process in terms of reaction kinetics, solid phase nucleation and growth processes. Hence, the solids obtained by chemical precipitation have a wide particle size distribution in addition to uncontrolled particle morphology along with agglomeration. To obtain nanoparticles with a narrow size distribution, the necessary requirements are a high degree of supersaturation, a uniform spatial concentration distribution inside a reactor and a uniform growth time for all the particles/crystals. Other commonly used solution method for the synthesis is the co-precipitation method. It produces a 'mixed' precipitate comprising two or more insoluble species that are simultaneously removed from solution. The precursors used in this method are mostly inorganic salts – nitrate, chloride, sulphate, *etc.* – that are dissolved in water or any other suitable medium to form a homogeneous solution with clusters of ions. After precipitation, the solid mass is collected, washed and gradually dried by heating to the boiling point of the medium [36]. The advantages of co-precipitation reactions include homogeneity of component distribution, low cost, relatively low reaction temperature and fine and uniform particle size with weakly agglomerated particles. The reactions are highly susceptible to the conditions. Moreover, control over the stoichiometry of the precursors is rather difficult to achieve, as the metal ions are incompletely precipitated. Additionally, the co-precipitation reactions are not suited for some systems such as amphoteric systems.

γ-Irradiation Method

γ-Irradiation is one of the effective methods for the synthesis of materials. It is extensively used in the preparation of nanocrystalline metals, alloys and polymer/metal nanocomposites. Among these materials, semiconducting chalcogenide/polymer nanocomposites have been sources of more attention. γ-Irradiation offers a means by which polymerization of monomers and formation of inorganic nanoparticles take place simultaneously. This leads to fabrication of inorganic/polymer nanocomposites. However, it is only quite recently that γ-irradiation has been applied to the synthesis of chalcogenide/polymer nanocomposites [37, 38]. The ease and versatility of this synthetic approach facilitate the development of functional nanomaterials.

Microwave Synthesis

Microwave irradiation has been used in the synthesis of inorganic nanoparticles and is now showing growth in its application to materials science as well. Compared with the conventional heating methods, microwave assisted heating presents a more rapid environment for the formation of nanoparticles. This is because of the fast and homogeneous heating effects of microwave irradiation. Hence, this method has the advantages of short reaction time, high energy efficiency and the ability to induce the formation of particles with small size, narrow size distribution and high purity. In the past few years, microwave assisted heating has been applied in the soft chemical synthesis of various nanocrystalline metal chalcogenides and also presents promising trends for future development [39 - 41].

SYNTHESIS OF THIN FILMS OF CHALCOGENIDES

Any sample with one of its dimensions very much less than the other two may be termed as thin film. In general, thin films are thin material layers ranging from fractions of nanometers to micrometers. The various properties, such as optical, thermal, electrical, magnetic, morphological *etc.* of thin films can be controlled by the method of deposition, substrate material and temperature, rate and angle of deposition, source-substrate distance and ambient atmosphere. Specific properties, such as optical reflection/transmission, hardness, adhesion, porosity, conductivity/resistivity, chemical inertness towards corrosive environments and stability with respect to temperature, stoichiometry *etc.* can be taken into consideration while synthesizing thin films according to desired applications.

Choice of Substrate

Substrate, also called wafer, is usually a planar solid substance on to which a layer of another substance is applied, and on to which the latter adheres. A substrate should possess qualities such as low cost, atomically smooth surface, inertness to chemicals used in processing, high thermal conductivity, high mechanical strength, high recrystallization temperature, coefficient of thermal expansion similar to that of deposited film, high thermal shock resistance so that it does not break due to soldering and excellent stripping and cutting properties so that it does not break while cutting.

It can be a thin slice of any material *viz.* silicon, silicon dioxide, aluminium oxide, sapphire, metal plate, glass, quartz, polymer sheet *etc.* Among these, glass substrates are widely used because of ease of availability and having amorphous structure, low cost, chemical inertness, high transmission capabilities in the visible region and poor thermal and electrical conductivities.

Cleaning and Drying of Substrate

Dust and various contaminants on the surface of substrate hinder proper adherence of the films to the substrate and subsequent reproducibility of the films. It is, therefore, important to duly clean the substrates prior to deposition of films. For the purpose of cleaning, the glass substrates are first dipped in chromic acid for few hours. These are then washed with liquid detergent using de-ionized water. The glass slides are finally cleaned in a digital ultrasonic cleaner (Fig. **1**) according to the following steps:

• De-ionized water, which is free from organic molecules, is poured into ultrasonic cleaner and the glass substrates are placed in it. These are subjected to ultrasonic agitation for 10-15 minutes. This de-ionized water is replaced by fresh one and the process is repeated.

Fig. (1). Ultrasonic cleaner (Model no. LMUC-2 from LABMAN).

• De-ionized water is replaced by methanol and the glass slides are ultrasonically agitated.

• Hereafter, methanol is replaced by acetone and the glass slides are again subjected to ultrasonic agitation for further 10-15 minutes.

The substrate is then held quasi-vertically with the help of forceps and its one face is dried by blowing heated air from hot-air gun so that no drops or streaks of any chemical or water are left on the substrate. The other face – front face on which the film is to be deposited – is dried in a similar manner. Rapid drying prevents stains on the glass surface which are formed due to re-deposition of impurities that water absorbs from the surrounding air. Thus, duly degreased and cleaned surface of the substrate gets ready for deposition of uniform film.

Methods of thin Film Deposition

Deposition of thin films means obtaining thin film of a desired material on the surface of substrate with proper adherence. Growth techniques have a defining role in controlling the properties of thin films . Structural defects formed during deposition are also significant for studying luminescence related properties. Chalcogenides have low phonon energy, extended infrared transparency, high refractive index and high photosensitivity [13]. Chalcogenide thin films have evoked increasing interest due to their exceptional properties, which are remarkably different from those of bulk materials [42]. The deposition techniques for synthesizing thin films are broadly classified into two categories:

• Physical Deposition methods

• Chemical Deposition methods

Physical Deposition Methods

Physical deposition methods mainly include thermal evaporation, sputtering, pulsed laser deposition, molecular beam epitaxy, liquid phase epitaxy *etc.* All these methods require maintenance of vacuum of 10^{-6} Torr or more [43]. Physical Vapour Deposition (PVD) is a process by which thin film of a material is deposited on a substrate according to the following sequence of steps:

• material to be deposited is converted into vapour by physical means *i.e.* heating

• vapour is transported across a region of low pressure from its source to the substrate

• vapour undergoes condensation on the substrate to form the thin film

Thermal Evaporation

This method has been used extensively for synthesizing chalcogenide thin films [44]. It has been found to be more suitable and useful due to high deposition rate, low consumption of material and economical way of deposition vitroprocess. In our group, thin films have been synthesized using thermal evaporation method. Vacuum coating unit Model No. 12A4D from HINDHIVAC Company, Bangalore, India (Fig. **2**) was used for depositing the thin films.

Thin films are synthesized by evaporation from a hot source on to a substrate. A sample of material to be deposited is placed in a molybdenum crucible/boat and the deposition chamber is evacuated up to ~ 10^{-6} Torr. The high-vacuum environment ensures that the vapourized atoms or molecules are transported to the substrate with minimal collision or interference from other gas atoms or molecules. The crucible/boat is then heated using a tungsten filament (resistive evaporation) or with an electron beam (e-beam evaporation) [45]. Evaporation can also be achieved by heating the source material with RF energy. This technique employs an RF induction heating coil that surrounds the crucible/boat containing the source. This method of evaporation is known as inductive heating evaporation. The evaporation technique is advantageous due to high film deposition rates, less substrate surface damage from impinging atoms as the film is being formed and the excellent purity of the film due to high vacuum conditions used during evaporation. However, the control of film composition is difficult as compared to that in sputtering and in-situ cleaning of substrate surfaces is not possible.

Fig. (2). Thermal evaporation coating unit in our laboratory. (Model No. 12A4D from HINDHIVAC Company, Bangalore, India).

Sputtering

Sputtering is a mechanism in which atoms are dislodged from the surface of a material as a result of collision with high-energy particles. The ejected atoms or molecules then condense on the substrate to form the thin film [46, 47]. The following sequence of steps take place while depositing films through sputtering:

• Ions are generated and directed at a target material

• The ions sputter atoms from the target

• The sputtered atoms get transported to the substrate through a region of reduced pressure

• The sputtered atoms condense on the substrate, forming a thin film.

Sputtering offers advantages over other physical deposition methods *viz.* easy control on film thickness by fixing the operating parameters and/or by adjusting the deposition time and deposition from large-size targets which simplifies the deposition of thin films with uniform thickness over large wafers. As compared to deposition through evaporation, sputtering can have better control of the material composition as well as other film properties. Sputter-cleaning of the substrate in vacuum prior to film deposition can be done. However, in sputtering high capital expenses are required. It is a slow process and some materials such as organic solids can get easily degraded by ionic bombardment.

Pulsed Laser Deposition (PLD)

This technique utilizes a high energy pulsed laser as an external power source for ablating the source or target material [48]. The interaction is short and intense and induces ablation *via* a cascade of complex events. A high energy pulsed laser – Nd:YAG or UV excimer – with intensity ~ $10^8 - 10^9$ W/cm^2 is placed outside the vacuum chamber. The beam is focused on to the target surface where it is absorbed in small volume, The absorbed energy density is sufficient to break any chemical bond between the molecules. So, a high pressure gas is produced in the surface layer. As a result of pressure gradient, a supersonic jet is ejected normal to the target surface. The particle cloud then absorbs a large amount of energy from the laser beam thus producing an expansion of hot plasma through the deposition chamber. Some hundreds of laser pulses falling on the target ablate the material, a plume is produced and the material condenses on the substrate and a thin film is formed. Under proper process parameters, the film grows epitaxially. The stoichiometry of the film is replica of that of the target. Deep surface pitting as a result of repetitive ablation can be avoided by beam rastering or target rotation or a combination of both. The separation between substrate and target, typically few meters, is such that the substrate just touches the tip of the ablation plume. This technique has the advantages of multilayer growth, control over crystallinity and maintenance of stoichiometry, but is limited to small area of substrates.

Molecular Beam Epitaxy (MBE)

It is a method that is capable of depositing layers comparable to atomic thickness. For this, a beam of molecules is created. The beam refers to evaporated atoms, which do not interact with one another until they are deposited on the substrate. MBE systems require high vacuum to work. So, a very low gas pressure from the outside world is maintained. Ultrapure elements are heated in effusion cells. These get hotter and hotter until they begin to sublimate. These elements start reacting with one another once they re-condense on the substrate. Crystal growth is measured by using a Reflection High Energy Electron Diffraction (RHEED) system. This technique brands the surface of crystal during epitaxy process. The crystal thickness is carefully controlled using computers to open and close the shutters in front of each furnace. This controlling method assures that different multiple layers are easily crafted into a variety of intricate layers. Molecular beam epitaxy is an elegant way to engineer crystals due to its excellent accuracy and precision. It is very important for the semiconductor industry. Used to grow and develop high purity crystals with very precise dimensions, MBE systems take excess molecules/atoms and allow them to condense on to a substrate. There they begin to form ultra-thin layers of crystal. This technique is optimal in the design of thin films, semiconductor lasers, solar cells, *etc* [49].

Chemical Deposition Methods

These methods primarily include chemical vapour deposition, chemical bath deposition, sol-gel spin coating, spray pyrolysis, electrochemical deposition, *etc.*

Chemical Vapour Deposition (CVD)

In this method, chemicals in the gas phase react and condense on the surface of the substrate so as to create high purity solid materials. Thus, a solid thin film is deposited from a gaseous phase as an outcome of chemical reaction. CVD starts when the substrate is exposed to one or more precursor gases (often diluted in carrier gases) and delivered into the reaction chamber at an ambient temperature. As soon as these gasses come into contact with a heated substrate, they react and/or decompose on the substrate surface forming a solid phase. Gas flow then removes all by-products through the reaction chamber. Coatings produced by CVD are by and large fine grained, impervious, highly pure and harder than other similarly conceived materials. These coatings are deposited at the rate of a few microns per hour. The films obtained are highly pure (99.999%) and are of high density. CVD, a versatile method used in a wide variety of fields, can be performed with any element or compound and many parts can be coated at the same time, making it economical. Due to its ability to produce components that are difficult to be made from conventional techniques, it is used in producing dense structural parts. Besides fabrication of solid state electronic devices, other applications include production of semiconductors, coatings, catalysts, optical fibers, composites, nanomachines, *etc* [50, 51].

Chemical Bath Deposition (CBD)

It is also known as chemical synthesis or solution growth technique. It is based on the formation of a solid phase from a solution and involves two steps – nucleation and particle growth. The method uses a controlled chemical reaction to deposit a thin film by precipitation. The substrate is immersed in a solution which contains the precursors. The growth of thin films strongly depends on growth conditions such as duration of deposition, composition and temperature of the solution. It also depends on the topographical and chemical nature of substrate. A number of physico-chemical factors *viz.* bath temperature, pH of the solution, molarity of concentration, duration, solubility product, supersaturation, type of precipitation *etc.* control the growth of deposit under a specified set of reaction conditions [52]. Uniform, stable, adherent and hard films are synthesized with good reproducibility by this simple process. In its simplest form, it requires only solution containers and substrate mounting devices. The main drawback is the wastage of solution after every deposition.

Sol-gel Spin Coating

In this process, the substrate spins around an axis perpendicular to the coating area. There are four distinct stages:

i) Initially, a controlled amount of the precursor is poured over the substrate. This wets the substrate uniformly. If need arises, sub-micron filter can be used to eliminate larger particles from the precursor.

ii) This step involves spinning of substrate with the desired rotation speed to remove the excess fluid. The top of the fluid layer exerts inertia while the substrate rotates at a faster speed. A twisting motion results due to these two forces. Thus, spiral vortices are formed. Generally, the precursor is kept thin enough so that it keeps co-rotating with the substrate and so, any thickness difference does not take place. Finally, the substrate reaches its desired speed. The viscous shear drag is then exactly balanced by rotational acceleration.

iii) In this stage, viscous forces are dominant with regard to thinning behaviour of the fluid. The thinning of fluid results in the formation of uniform films. Sometimes, edge effects are also seen. The rotating fluid tends to flow uniformly outwards. Drops are formed at the edges if the fluid is in excess. Thus, in some cases, thickness at the ends of the film is observed to be slightly greater than that at the central portion of the substrate. This leads to edge effect.

iv) Finally, the fluid starts evaporating and starts dominating its own thinning process. It is the step where solvent phase gets removed and the sol is converted into dense ceramic. The high speed of rotating fluid tends to increase its temperature and leads to its evaporation. Thus, the viscosity of the remaining solution increases. This is the 'gel-state' of the coating. Although stages iii) and

iv) *i.e.* viscous flow and evaporation occur simultaneously, the viscous flow effect dominates initially and evaporation dominates later. The sol is kept in an air tight flask to maintain its viscosity; otherwise, it can convert into gel and cannot be used for deposition of films. The factors that affect thickness of the film are molar concentration, viscosity of the coating solution, rotation rate, time duration and annealing temperature. Gel coated films are found to be porous as and when they are heated [44, 53].

Spray Pyrolysis

It is essentially a thermally stimulated reaction between clusters of liquid/vapour atoms of different chemical species. This technique usually involves spraying precursor solution containing salts of the constituent atoms of the desired compound on a substrate which is maintained at elevated temperatures. Once the sprayed droplets reach the hot substrate, they undergo pyrolytic decomposition. Thus, single crystals or crystallites of the product are formed. The other volatile byproducts as well as excess solvent escape through the vapour phase. The hot substrates provide the necessary thermal energy for deposition and subsequent recombination of the species and recrystallization of the crystallites. The optimization of chemical solution into a spray of fine droplets is effected by the spray nozzle. This is done with the help of a carrier gas which is not involved in the pyrolytic reaction. Large area uniform coverage of the substrate is achieved by scanning either or both the spray head and the substrate employing electro-mechanical arrangements. The chemicals used for spray pyrolysis must satisfy the following conditions:

• The desired thin film be obtained as a consequence of thermally activated reaction between the various species/complexes dissolved in spray solutions.

• The remainder constituents of the chemicals, including the carrier liquid, be volatile at the pyrolysis temperature.

The above conditions can be satisfied by a number of combinations of chemicals for a given thin film material [54 - 56]. Deposition parameters tend to vary the structure and morphology of the films. Growth of thin films is determined by the nature of substrate & chemical, concentration of spray solution & its additives and spray parameters. The spray deposited films are generally adherent, pinhole free, mechanically hard and stable with time and temperature. The topography of the films is mostly rough and dependent on spray conditions.

OTHER METHODS

Apart from the aforementioned methods, synthesis of chalcogenide thin films is also done by using arrested precipitation technique [57], photoelectrochemical deposition [58, 59], Metal Organic Chemical Vapour Deposition (MOCVD) [60], electrodeposition [61 - 65], Successive Ionic Layer Adsorption and Reaction (SILAR) [66, 67], close-spaced vacuum sublimation [68], solvothermal synthesis [69, 70], hydrothermal synthesis [71, 72], hot wall deposition [73], *etc.*

ANNEALING

Annealing involves heating, maintaining a suitable temperature and cooling. As a result, diffusion of atoms occurs within a solid material which helps material progress towards its equilibrium state. The three stages of annealing process that occur as the temperature of the material is increased are: recovery, recrystallization and grain growth. The first stage results in softening of the metal by removing primarily linear defects called dislocations and the internal stresses they cause. Recovery occurs at the lower temperature stage of all the annealing processes and before the appearance of new strain-free grains. The second stage of recrystallization is the one where new strain-free grains nucleate. These then grow to replace those deformed by internal stresses. If annealing continues, the third stage of grain growth is found to occur. The microstructure then starts to coarsen. This causes the material to lose a substantial part of its original strength which can, however, be regained with hardening [74].

Annealing has played a very significant role in the development of Se-Te and Te-Se based samples in our laboratory. All the reported bulk as well as thin film samples have been subjected to this process and their results put forth for analysis. Effects of annealing on various properties of chalcogenides in both bulk and thin film forms have been reported [75].

SAMPLES PREPARED IN OUR LABORATORY

Bulk samples $Se_{90-x}Te_{10}Cu_x$ (x = 0, 10, 20) are synthesized by melt quenching technique. 99.999% pure constituent elements – Se, Te and Cu – purchased from Sigma Aldrich were weighed, as per requirement, in proportion of their respective atomic weight percentages using electronic balance Citizen CY 64, which is capable of measuring up to three decimal places of a gram. The sample-wise respective weighed constituents were then put in three separate quartz ampoules each having length ~ 8 cm and internal diameter ~ 12 mm. The ampoules were already cleaned by rinsing with soap solution, acetone and methanol for removing the impurities. Vacuum of ~ 10^{-3} Torr was created by a rotary pump in the ampoules so as to avoid any reaction of materials with oxygen at high temperature. During the process of evacuation itself, each ampoule was sealed thermally with the help of Oxygen-Liquefied Petroleum Gas (LPG) flame torch. The sealed ampoules were then placed in a microprocessor-controlled programmable muffle furnace. Heating was done at a rate of 4°C/min from room temperature to 1000°C and the ampoules were kept at this temperature for 10 hours. During heating, the ampoules were frequently rocked to ensure homogenization and proper mixing of the melt. After 10 hours, the red hot ampoules were taken out from the furnace and the melts were quenched by dropping the ampoules immediately into ice-cold water for the materials to go into glassy state. The quartz ampoules were then broken and the quenched samples were recovered. The three solid samples so obtained were then crushed separately with the help of mortar/pestle and sieved with the help of sieve-mesh. The coarse part was again crushed and then sieved. The process of crushing and sieving was followed repeatedly till all the three bulk samples *viz.* $Se_{90}Te_{10}$, $Se_{80}Te_{10}Cu_{10}$ and $Se_{70}Te_{10}Cu_{20}$ were obtained in the form of fine powder. Similarly, other samples which included quaternary samples were also prepared. Similarly, other samples having excess of Te *i.e.* Te-Se based samples have also been prepared in our lab. Thin film samples were obtained using thermal evaporation [76]. Annealing was done at the rate of 5°C/min from room temperature up to 125°C and this temperature was maintained for two hours. The samples were then allowed to cool back naturally to room temperature. Various properties such as structural, morphological, thermal, optical and photoluminescence have been studied and reported.

CONCLUSION

Chalcogenides – sulphides, selenides and tellurides – consisting of at least one chalcogen element (S, Se, Te) and at least one more electropositive element find potential applications in thin film solar cells, photoelectrochemical cell, heterojunction solar cells, photodetectors, photoconductive cell, *etc.* Chalcogenides are also used as high performance counter electrodes in dye sensitized solar cells, catalysts, optical coatings and photoelectrodes. In bulk form, binary, ternary or quaternary chalcogenides can be synthesized by melt quenching technique, thermolysis/thermal decomposition, chemical precipitation or microwave synthesis. Chalcogenide thin films can be obtained by various physical methods *viz.* thermal evaporation, sputtering, pulsed laser deposition, molecular beam epitaxy and chemical methods *viz.* chemical vapour deposition, chemical bath deposition, sol-gel spin coating, spray pyrolysis, *etc.* Annealing plays a significant role in modifying the sample properties. We have published our research work regarding the synthesis and deposition of Se-Te-Cu and Se-T--Cu-Cd bulk and thin films, the references of which are given here.

CONSENT FOR PUBLICATION

Not applicable.

CONFLICT OF INTEREST

The authors declare no conflict of interest, financial or otherwise.

ACKNOWLEDGEMENTS

Financial assistance from UGC, New Delhi vide project F. No. 42-773/2013(SR) is gratefully acknowledged. Authors are also thankful to U.P. State Government for providing XRD facility at the Department of Physics, University of Lucknow, Lucknow through Centre of Excellence Scheme.

REFERENCES

[1]   Zakai U. 2007. Design, Synthesis and Evaluation of Chalcogen interactions. Ph.D. Thesis, University of Arizona, 2007.

[2]   Sagadevan S, Chandraseelan E. Applications of Chalcogenide Glasses: An Overview. Int J Chemtech Res 2014; 6: 4682-6.

[3]   Shukla S, Kumar S. Photoconductivity and high-field effects in amorphous $Se_{83}Te_{15}Zn_2$ thin film. Bull Mater Sci 2011; 34(7): 1351-5.
[http://dx.doi.org/10.1007/s12034-011-0327-6]

[4]   Avrami M. Kinetics of Phase Change. II Transformation-Time Relations for Random Distribution of Nuclei. J Chem Phys 1940; 8(2): 212-24.
[http://dx.doi.org/10.1063/1.1750631]

[5]   Mahadevan S, Giridhar A, Singh AK. Calorimetric measurements on as-sb-se glasses. J Non-Cryst Solids 1986; 88(1): 11-34.
[http://dx.doi.org/10.1016/S0022-3093(86)80084-9]

[6]   Kumar R, Sharma P, Rangra VS. Kinetic studies of bulk $Se_{92}Te_{8-x}$ Sn $_x$ (x = 0, 1, 2, 3, 4 and 5) semiconducting glasses by DSC technique. J Therm Anal Calorim 2012; 109(1): 177-81.
[http://dx.doi.org/10.1007/s10973-011-1661-z]

[7]   Kumar R, Sharma P, Katyal SC, Sharma P, Rangra VS. A study of Sn addition on bonding arrangement of Se-Te alloys using far infrared transmission spectroscopy. J Appl Phys 2011; 110(1): 013505.
[http://dx.doi.org/10.1063/1.3603010]

[8]   Kumar K, Sharma P, Katyal SC, Thakur N. Effect of Bi addition on dc, ac conductivity and dielectric properties of $Te_{15}(Se_{100-x}Bi_x)_{85}$ glassy alloys. J Optoelectron Adv Mater 2011; 13: 371-6.

[9]   Suri N, Bindra KS, Thangaraj R. Transport properties of $Se_{80-x}Te_{20}Bi_x$ (0 < x < 1) system. J Non-Cryst Solids 2007; 353(18-21): 2079-83.
[http://dx.doi.org/10.1016/j.jnoncrysol.2007.01.071]

[10]   Suri N, Bindra KS, Kumar P, Thangaraj R. Calorimetric studies of $Se_{80-x}Te_{20}Bi_x$ bulk samples. J Non-Cryst Solids 2007; 353(13-15): 1264-7.

[http://dx.doi.org/10.1016/j.jnoncrysol.2006.10.056]

[11]   Suri N, Bindra KS, Thangaraj R. Electrical conduction and photoconduction in Se_{80-x} Te_{20} Bi_x thin films. J Phys Condens Matter 2006; 18(39): 9129-34.
[http://dx.doi.org/10.1088/0953-8984/18/39/038]

[12]   Kamboj MS, Thangaraj R. Calorimetric studies of bulk Se-Te-Pb glassy system. Eur Phys J Appl Phys 2003; 24(1): 33-6.
[http://dx.doi.org/10.1051/epjap:2003052]

[13]   Sharma K, Lal M, Kumar A, Goyal N. Investigation of optical constants and optical band gap for amorphous $Se_{70}Te_{30-x}Sb_x$ thin films. J Optoelectron Biomed Mater 2014; 6: 19-26.

[14]   Saraswat VK, Kishore V, Singh K, Saxena NS, Sharma TP. Electrical conductivity of Se-In chalcogenide glasses. Chalc Lett 2006; 3: 61-5.

[15]   Mikla VI, Mikla VV. Amorphous ChalcogenidesThe Past, Present and Future. Amsterdam: Elsevier 2012.

[16]   Practical Veterinary Diagnostic Imaging, 2nd Edition, West Sussex, United Kingdom: Wiley-Blackwell; john Wiley & Sons, 2012.

[17]   Pradeep P, Saxena NS, Saxena MP, Kumar A. Crystallization kinetics of Se-Te-Cd glasses. Phys Status Solidi 1996; 23: 156. [a].

[18]   Pawar SM, Moholkar AV, Rajpure KY, Bhosale CH. Photoelectrochemical investigations on electrochemically deposited CdSe and Fe-doped CdSe thin films. Sol Energy Mater Sol Cells 2008; 92(1): 45-9.
[http://dx.doi.org/10.1016/j.solmat.2007.08.011]

[19]   Shinde SK, Dubal DP, Ghodake GS, Fulari VJ. Morphological modulation of Mn:CdSe thin film and its enhanced electrochemical properties. J Electroanal Chem (Lausanne) 2014; 727: 179-83.
[http://dx.doi.org/10.1016/j.jelechem.2014.04.005]

[20]   Tian L, Yang H, Ding J, Li Q, Mu Y, Zhang Y. Synthesis of the wheat-like CdSe/CdTe thin film heterojunction and their photovoltaic applications. Curr Appl Phys 2014; 14(6): 881-5.
[http://dx.doi.org/10.1016/j.cap.2014.04.003]

[21]   Lou S, Zhou C, Wang H, *et al.* Annealing effects on the photoresponse properties of CdSe nanocrystal thin films. Mater Chem Phys 2011; 128(3): 483-8.
[http://dx.doi.org/10.1016/j.matchemphys.2011.03.035]

[22]   Faber MS, Park K, Cabán-Acevedo M, Santra PK, Jin S. Earth-Abundant Cobalt Pyrite (CoS_2) Thin Film on Glass as a Robust, High-Performance Counter Electrode for Quantum Dot-Sensitized Solar Cells. J Phys Chem Lett 2013; 4(11): 1843-9.
[http://dx.doi.org/10.1021/jz400642e] [PMID: 26283119]

[23]   Gohar S, Pathak R. Potentiostatic Electrochemical Preparation and Characterisation of Aluminium Containing Nickel Selenide. Orient J Chem 2013; 29(4): 1469-74.
[http://dx.doi.org/10.13005/ojc/290424]

[24]   Smotkin ES, Cervera-March S, Bard AJ, *et al.* Bipolar cadmium selenide/cobalt(II) sulfide semiconductor photoelectrode arrays for unassisted photolytic water splitting. J Phys Chem 1987; 91(1): 6-8.
[http://dx.doi.org/10.1021/j100285a003]

[25]   Labadie L, Vigreux-Bercovici C, Pradel A, Kern P, Arezki B, Broquin JE. M-lines characterization of selenide and telluride thick films for mid-infrared interferometry. Opt Express 2006; 14(18): 8459-69.
[http://dx.doi.org/10.1364/OE.14.008459] [PMID: 19529223]

[26]   Abe K, Takebe H, Morinaga K. Preparation and properties of Ge-Ga-S glasses for laser hosts. J Non-Cryst Solids 1997; 212(2-3): 143-50.
[http://dx.doi.org/10.1016/S0022-3093(96)00655-2]

[27] Kozyukhin SA, Popov AI, Voronkov EN. Influence of chalcogenide glasses electro physical parameters on threshold voltage for phase-change memory. Thin Solid Films 2010; 518(20): 5656-8.
[http://dx.doi.org/10.1016/j.tsf.2009.10.033]

[28] Hu J, Tarasov V, Carlie N, *et al.* Exploration of waveguide fabrication from thermally evaporated Ge–Sb–S glass films. Opt Mater 2008; 30(10): 1560-6.
[http://dx.doi.org/10.1016/j.optmat.2007.10.002]

[29] Maurugeon S, Boussard-Plédel C, Troles J, *et al.* Telluride Glass Step Index Fiber for the far Infrared. J Lightwave Technol 2010; 28: 3358-63.
[http://dx.doi.org/10.1109/JLT.2010.2087008]

[30] Kumar S, Singh K. Glass transition, thermal stability and glass-forming tendency of $Se_{90-x}Te_5Sn_5In_x$ multi-component chalcogenide glasses. Thermochim Acta 2012; 528: 32-7.
[http://dx.doi.org/10.1016/j.tca.2011.11.005]

[31] Asobe M, Itoh H, Miyazawa T, Kanamori T. Efficient and ultrafast all-optical switching using high Δn, small core chalcogenide glass fibre. Electron Lett 1993; 29(22): 1966-8.
[http://dx.doi.org/10.1049/el:19931309]

[32] Ta'eed VG, Fu L, Pelusi M, *et al.* Error free all optical wavelength conversion in highly nonlinear As-Se chalcogenide glass fiber. Opt Express 2006; 14(22): 10371-6.
[http://dx.doi.org/10.1364/OE.14.010371] [PMID: 19529435]

[33] Song KY, Abedin KS, Hotate K, González Herráez M, Thévenaz L. Highly efficient Brillouin slow and fast light using As_2Se_3 chalcogenide fiber. Opt Express 2006; 14(13): 5860-5.
[http://dx.doi.org/10.1364/OE.14.005860] [PMID: 19516755]

[34] Kumar A, Lal M, Sharma K, Tripathi SK, Goyal N. DC conduction and Meyer-Neldel Rule in $Se_{85-x}Te_{15}In$. J Non-Oxide Glasses 2013; 5: 27-31.

[35] Nogriya V, Dongre JK, Ramrakhiani M, Chandra BP. Electro-and photo-luminescence studies of CdS nanocrystals prepared by orgenometallic precursor. Chalcogenide Lett 2008; 5: 365-73.

[36] Srinivasa Rao B, Rajesh Kumar B, Rajagopal Reddy V, Subba Rao T. Preparation and characterization of CdS nanoparticles by chemical co-precipitation technique. Chalcogenide Lett 2011; 8: 177-85.

[37] Qiao Z, Xie Y, Li X, Wang C, Zhu Y, Qian Y. γ-Irradiation preparation and phase control of nanocrystalline CdS. J Mater Chem 1999; 9(3): 735-8.
[http://dx.doi.org/10.1039/a807757f]

[38] Ge X, Ni Y, Liu H, Ye Q, Zhang Z. γ-Irradiation preparation of cadmium selenide nano-particles in ethylenediamine system. Mater Res Bull 2001; 36(9): 1609-13.
[http://dx.doi.org/10.1016/S0025-5408(01)00638-9]

[39] Washington AL II, Strouse GF. Microwave synthesis of CdSe and CdTe nanocrystals in nonabsorbing alkanes. J Am Chem Soc 2008; 130(28): 8916-22.
[http://dx.doi.org/10.1021/ja711115r] [PMID: 18576624]

[40] Prasad N, Furniss D, Rowe HL, Miller CA, Gregory DH, Seddon AB. First time microwave synthesis of $As_{40}Se_{60}$ chalcogenide glass. J Non-Cryst Solids 2010; 356(41-42): 2134-45.
[http://dx.doi.org/10.1016/j.jnoncrysol.2010.08.006]

[41] Ding K, Lu H, Zhang Y, *et al.* Microwave synthesis of microstructured and nanostructured metal chalcogenides from elemental precursors in phosphonium ionic liquids. J Am Chem Soc 2014; 136(44): 15465-8.
[http://dx.doi.org/10.1021/ja508628q] [PMID: 25333207]

[42] Hirohata A, Moodera JS, Berera GP. Structural and electrical properties of InSe polycrystalline films and diode fabrication. Thin Solid Films 2006; 510(1-2): 247-50.
[http://dx.doi.org/10.1016/j.tsf.2005.12.202]

[43] David MK, Devadason S. A comparative study on the optical properties of multilayer CdSe/CdTe thin

film with single layer CdTe and CdSe films. Jounal of Nano-and Electronic Physics 2013; 5: 1-4.

[44]   Krbal M, Wagner T, Kohoutek T, Nemec P, Orava J, Frumar M. The comparison of Ag–As33S67 films prepared by thermal evaporation (TE), spin-coating (SC) and a pulsed laser deposition (PLD). J Phys Chem Solids 2007; 68(5-6): 953-7.
[http://dx.doi.org/10.1016/j.jpcs.2007.03.036]

[45]   Suthan Kissinger NJ, Suthagar J, Saravana Kumar B, Balasubramaniam T, Perumal K. Effect of substrate temperature on the structural and optical properties of nanocrystalline cadmium selenide thin films prepared by electron beam evaporation technique. Acta Phys Pol A 2010; 118(4): 623-8.
[http://dx.doi.org/10.12693/APhysPolA.118.623]

[46]   Campos-González E, Rodríguez-Fragoso P, Gonzalez de la Cruz G, Santoyo-Salazar J, Zelaya-Angel O. Synthesis of CdSe nanoparticles immersed in an organic matrix of amylopectin by means of rf sputtering. J Cryst Growth 2012; 338(1): 251-5.
[http://dx.doi.org/10.1016/j.jcrysgro.2011.10.046]

[47]   Nazabal V, Charpentier F, Adam JL, *et al.* Sputtering and pulsed laser deposition for near- and mid-infrared applications: a comparative study of $Ge_{25}Sb_{10}S_{65}$ and $Ge_{25}Sb_{10}Se_{65}$ amorphous thin films. Int J Appl Ceram Technol 2011; 8(5): 990-1000.
[http://dx.doi.org/10.1111/j.1744-7402.2010.02571.x]

[48]   Erazú M, Rocca J, Fontana M, Ureña A, Arcondo B, Pradel A. Raman spectroscopy of chalcogenide thin films prepared by PLD. J Alloys Compd 2010; 495(2): 642-5.
[http://dx.doi.org/10.1016/j.jallcom.2009.10.251]

[49]   Yang Q, Zhao J, Guan M, *et al.* Growth and annealing of zinc-blende CdSe thin films on GaAs (0 0 1) by molecular beam epitaxy. Appl Surf Sci 2011; 257: 9038-43.
[http://dx.doi.org/10.1016/j.apsusc.2011.05.096]

[50]   Monkman AP, Adams P. Optical and electronic properties of stretch-oriented solution-cast polyaniline films. Synth Met 1991; 40(1): 87-96.
[http://dx.doi.org/10.1016/0379-6779(91)91491-R]

[51]   Panneerselvam A, Malik MA, Afzaal M, O'Brien P, Helliwell M. The chemical vapor deposition of nickel phosphide or selenide thin films from a single precursor. J Am Chem Soc 2008; 130(8): 2420-1.
[http://dx.doi.org/10.1021/ja078202j] [PMID: 18237172]

[52]   Chavhan S, Sharma R. Growth, structural and optical properties of non-stoichiometric $CuIn(S_{1-x}Se_x)2$ thin films deposited by solution growth technique for photovoltaic application. J Phys Chem Solids 2006; 67(4): 767-73.
[http://dx.doi.org/10.1016/j.jpcs.2005.11.013]

[53]   Tsay CY, Cheng HC, Chen CY, Yang KJ, Lin CK. The properties of transparent semiconductor Zn1−Ti O thin films prepared by the sol–gel method. Thin Solid Films 2009; 518(5): 1603-6.
[http://dx.doi.org/10.1016/j.tsf.2009.09.054]

[54]   Raturi AK, Thangaraj R, Sharma AK, Tripathi BB, Agnihotri OP. Structural, optical and photoconducting properties of sprayed CdSe films. Thin Solid Films 1982; 91(1): 55-64.
[http://dx.doi.org/10.1016/0040-6090(82)90123-7]

[55]   Kamoun N, Bouzouita H, Rezig B. Fabrication and characterization of Cu_2ZnSnS_4 thin films deposited by spray pyrolysis technique. Thin Solid Films 2007; 515(15): 5949-52.
[http://dx.doi.org/10.1016/j.tsf.2006.12.144]

[56]   Yadav AA, Barote MA, Masumdar EU. Studies on cadmium selenide (CdSe) thin films deposited by spray pyrolysis. Mater Chem Phys 2010; 121(1-2): 53-7.
[http://dx.doi.org/10.1016/j.matchemphys.2009.12.039]

[57]   Bhosale PN, Mane RM, Ghanwat VB, Kondalkar VV, Mane SR, Khot KV. Arrested Precipitation Technique for Synthesis of Chalcogenide and Oxide Thin Films. J Adv Chem Eng 2015; 5: e104.

[58]   Chate PA, Patil SS, Patil JS, Sathe DJ, Hankare PP. Synthesis, optoelectronic properties and

photoelectrochemical performance of CdS thin films. Physica B 2013; 411: 118-21.
[http://dx.doi.org/10.1016/j.physb.2012.11.032]

[59]   Pawar SM, Moholkar AV, Rajpure KY, Bhosale CH. Electrosynthesis and characterization of CdSe thin films: Optimization of preparative parameters by photoelectrochemical technique. J Phys Chem Solids 2006; 67(11): 2386-91.
[http://dx.doi.org/10.1016/j.jpcs.2006.06.015]

[60]   Chae DY, Seo KW, Lee SS, Yoon SH, Shim IW. CdSe thin films grown by MOCVD method using new single-source precursors. Bull Korean Chem Soc 2006; 27(5): 762-4.
[http://dx.doi.org/10.5012/bkcs.2006.27.5.762]

[61]   Kuranouchi S, Nakazawa T. Study of one-step electrodeposition condition for preparation of CuIn(Se, S)$_2$ thin films. Sol Energy Mater Sol Cells 1998; 50(1-4): 31-6.
[http://dx.doi.org/10.1016/S0927-0248(97)00986-0]

[62]   Henríquez R, Badán A, Grez P, *et al.* Electrodeposition of nanocrystalline CdSe thin films from dimethyl sulfoxide solution: Nucleation and growth mechanism, structural and optical studies. Electrochim Acta 2011; 56(13): 4895-901.
[http://dx.doi.org/10.1016/j.electacta.2011.02.113]

[63]   Zainal Z, Saravanan N, Mien HL. Electrodeposition of nickel selenide thin films in the presence of triethanolamine as a complexing agent. J Mater Sci Mater Electron 2005; 16(2): 111-7.
[http://dx.doi.org/10.1007/s10854-005-6460-5]

[64]   Manivannan R, Victoria SN. Preparation of chalcogenide thin films using electrodeposition method for solar cell applications – A review. Sol Energy 2018; 173: 1144-57.
[http://dx.doi.org/10.1016/j.solener.2018.08.057]

[65]   Carim AI, Saadi FH, Soriaga MP, Lewis NS. Electrocatalysis of the hydrogen-evolution reaction by electrodeposited amorphous cobalt selenide films. J Mater Chem A Mater Energy Sustain 2014; 2(34): 13835-9.
[http://dx.doi.org/10.1039/C4TA02611J]

[66]   Pathan HM, Sankapal BR, Desai JD, Lokhande CD. Preparation and characterization of nanocrystalline CdSe thin films deposited by SILAR method. Mater Chem Phys 2003; 78(1): 11-4.
[http://dx.doi.org/10.1016/S0254-0584(02)00198-0]

[67]   Pathan HM, Lokhande CD. Deposition of metal chalcogenide thin films by successive ionic layer adsorption and reaction (SILAR) method. Bull Mater Sci 2004; 27(2): 85-111.
[http://dx.doi.org/10.1007/BF02708491]

[68]   Gnatenko YP, Opanasyuk AS, Ivashchenko MM, Bukivskij PM, Faryna IO. Study of the correlation between structural and photoluminescence properties of CdSe thin films deposited by close-spaced vacuum sublimation. Mater Sci Semicond Process 2014; 26: 663-8.
[http://dx.doi.org/10.1016/j.mssp.2014.06.013]

[69]   Han ZH, Yu SH, Li YP, *et al.* Convenient solvothermal synthesis and phase control of nickel selenides with different morphologies. Chem Mater 1999; 11(9): 2302-4.
[http://dx.doi.org/10.1021/cm990198q]

[70]   Zhao JF, Song JM, Liu CC, *et al.* Graphene-like cobalt selenide nanostructures: template-free solvothermal synthesis, characterization and wastewater treatment. CrystEngComm 2011; 13(19): 5681-4.
[http://dx.doi.org/10.1039/c1ce05323j]

[71]   Zhuang Z, Peng Q, Zhuang J, Wang X, Li Y. Controlled hydrothermal synthesis and structural characterization of a nickel selenide series. Chemistry 2006; 12(1): 211-7.
[http://dx.doi.org/10.1002/chem.200500724] [PMID: 16259035]

[72]   Sobhani A, Salavati-Niasari M, Davar F. Shape control of nickel selenides synthesized by a simple hydrothermal reduction process. Polyhedron 2012; 31(1): 210-6.

[http://dx.doi.org/10.1016/j.poly.2011.09.017]

[73]   Velumani S, Narayandass SK, Mangalaraj D. Structural characterization of hot wall deposited cadmium selenide thin films. Semicond Sci Technol 1998; 13(9): 1016-24.
[http://dx.doi.org/10.1088/0268-1242/13/9/009]

[74]   Verhoeven J. Fundamentals of Physical Metallurgy. New York: Wiley, 1975.

[75]   Chandra P, Verma AK, Shukla RK, Srivastava A. Effects of annealing on structural and optical properties of $Se_{90-x}Te_{10}Cu_x$ (x = 0, 10, 20) thin films. Int j pure appl phys 2017; 13: 459-69.

[76]   Chandra P, Verma AK, Shukla RK, Srivastava A. Structural, morphological and optical characterization of as-deposited and annealed quaternary $Se_{50}Cd_{20}Cu_{20}Te_{10}$ chalcogenide thin film. Int J Mater Sci 2017; 12: 531-9.

CHAPTER 2

Recent Advances in Chalcogenide Glasses and their Applications

Horesh Kumar[1,*] and **Achchhe Lal Saroj**[1]

[1] *Department of Physics, Institute of Science, Banaras Hindu University, Varanasi-221005, UP, India*

Abstract: During the last two decades, by using a combination of both chalcogens (sulfur (S), selenium (Se), tellurium (Te), and polonium (Po)) and other elements like silicon (Si) and germanium (Ge), a huge number of chalcogenide glasses (ChGs) were prepared and studied. Compared to oxide-based glassy materials, ChGs have unique properties and functionalities which make them suitable for photonic applications. These materials are transparent in nature from the visible to the near-infrared region and can be used for the preparation of optical and electronic devices like ChG fibers, optical switches, sensors, and phase change memorizers. This chapter deals with some basics of ChGs, preparation techniques and a review of the latest technological development. The structural properties, optical properties, thermal and electrical properties of ChGs have been discussed. The physical aging effect has been explored. In the second part of this chapter, the applications of ChGs especially in dye sensitized solar cells (DSSCs), semiconductors, electrical memories and phase change memories have been discussed.

Keywords: Chalcogenide Glasses, Dielectric Properties, Electrical Conductivity, Melt-quench Technique, Optical Properties.

INTRODUCTION

Generally, materials are classified as solid, liquid, and gas. Solids have subcategories as crystalline and amorphous solids. Amorphous solids are considered distorted crystalline solids. X-ray scans of such solid comprise broad halos instead of characteristic sharp peaks as in the case of crystalline materials. A subclass of amorphous materials is Glass, which is the result of cooling from a normal cooling state to a rigid state without crystallization [1 - 3]. During the transition from a supercooled state to a glassy state, first-order thermo dynamical

* **Corresponding author Horesh Kumar:** Department of Physics, Institute of Science, Banaras Hindu University, Varanasi-221005, UP, India; E-mail: kumarhoresh@gmail.com

Arti Srivastava, Mridula Tripathi, Kalpana Awasthi and Subhash Banerjee (Eds.)

properties like volume(V), entropy(S) and heat content(H) remain continuous while second-order thermodynamical quantities like specific heat and thermal expansivity become discontinuous. Glasses have affected our life immensely as an optical material, insulating, and thermoelectric material. Depending upon the main constitutes of the glass, glasses are divided into three classes namely Fluoride-based glass, Heavy metal oxide (HMO) and Chalcogenide glasses. Chalcogenide glasses are inorganic glassy materials that contain as a key constituent one or more of the chalcogen elements (S, Se and Te except oxygen). These glasses are equipped with striking optical, thermal, electronic and electrical properties such as transmission extended into the infrared region, a higher refractive index, exhibit optical nonlinearities, optically and electrically induced memory effect and conductivity (photoconductivity). Obeying to these properties, they have applications in infrared devices, night-vision devices, solar cells/Dye-sensitized solar cells (DSSCs), waveguides, optical fibers sensors and lasers [4 - 6].

Structure of Chalcogenide Glasses (ChGs)

The glasses are non-crystalline (or amorphous) substances that lack a long-range order of arrangement of atoms in contrary to crystalline materials in which the structure can be described in terms of a periodically-continued, finite-sized unit cell. The structure of a glass consists of a bonding structure, called a normal bonding structure, which is topologically the same as the crystalline materials plus a defective structure [7]. The normal bonding structure is further divided into the short (≥ 0.5 nm) and the medium-range (0.5−3 nm) structures. The defective structure is similar to a defect in practical crystals like dangling and wrong bonds; however, defects such as dislocations and stacking faults do not exist in glassy materials. Short range order is determined by the atoms of the first coordination sphere and partially by the atoms of the second coordination sphere which are covalently bonded to the central atom [8]. Umbrella of short-range order encompasses the parameters like the number of nearest neighbors (first coordination number), their types, the distance between them and the central atom (radius of the first coordination sphere), their angle positions with respect to the central atom defined by bond angles (valency angles). The regular distribution of dihedral angles for a distance of about 10 atoms [9] determines the medium range order. In chalcogenide glasses, weak Van der Waals force between structural units determines medium-range order.

It is extremely difficult to determine the structure of glassy materials by a single technique (X-ray diffraction), static disorder in bond lengths and bond angles leads to a broader radial distribution function and hence the use of complementary techniques is unavoidable. Notable techniques are vibrational spectroscopic techniques such as IR absorption and Raman scattering, extended X-ray absorption fine-structure spectroscopy (EXAFS), magic-angle spinning nuclear magnetic resonance (MASNMR) and related NMR techniques, Scanning tunneling and atomic force microscopy (AFM). The structure of multi-component glasses has been explored using EXAFS. Direct images with diffraction patterns are facilitated by Transmission electron microscopy (TEM) [10]. Surface morphology is usually determined with the help of SEM (Scanning tunneling microscopy) and AFM [11]. X-ray photoelectron spectroscopy (XPS) is useful in determining the valences of constituent atoms [12].

Glassy selenium (g-Se) is the most prevalent form of non-crystalline selenium. There exist various forms (rings and chains) of selenium in the glassy state. These forms of selenium differ from each other in terms of ratio, size, form and mutual packing of structural units. The proportion of chains [Trans-coupling configuration, (Fig. **1b**)] and rings [*Cis*-coupling configuration, (Fig. **1a**)] in a particular sample depends on the method of preparation. Kaplow and Averbach [13] have studied the vitreous selenium and concluded that its structure consists of 95% selenium rings (8-membered) and 5% chain-like (strongly distorted) units. Some researchers [14] prepared films of g-Se, which consisted of only ring molecules Se_8 (amorphous analog of monoclinic selenium) or of only chain molecules Se_n (amorphous analog of trigonal selenium). Several models have been suggested for describing the structure of selenium. Richter and Breiting [15] put forward a model in which flat zigzag chains having a zero value of dihedral angle are proposed as structural units. Malaurent and Dixmier suggested a model in which any value of it permitted [16] is known as the 'free rotation chain model' (Fig. **2a**). Another model, known as the disordered chain model, was given by Lucovsky and Galeener [9] in which the changes in the sign of dihedral angle are allowed but changes in the value of dihedral angle are barred (Fig. **2b**).

Electronic Structure of Chalcogenide Glasses

We do not have Bloch-like wave functions for disordered solids. The disorder leads to the localization of electron wave function to a special extent Δr, which satisfies the uncertainty relation, $\Delta r.\Delta p \sim 1$ [17, 18]. Bonding and energy levels of atoms give rise to molecular orbitals. Overlapping of molecular orbitals leads to broadening of individual levels and hence, splitting into bands. DOS (Density of states) in glass is similar to the corresponding crystal DOS because both have short-range order. Therefore, band gap energies (E_g) and also optical absorptions at $h\omega > E_g$ are roughly the same. However, band edge structure is a modified medium-range structure (in crystalline materials, band edges are sharp) [19] and the modifications may be different in the conduction and the valence band. In chalcogenides, anti-bonding molecular orbitals form the conduction and lone-pair electrons states that form the valance band which will reflect structure differently. Defects also have an effect on band structure. For example, dangling bonds produce mid-gap states. These states govern optical absorption and photoluminescence. The configurational disorder leads to a fluctuation in the potential faced by carriers. With this fluctuating potential, we have localized states in the tail of DOS. Since electrons cannot diffuse at T= 0°K in localized states, they can contribute to the conductivity by phonon-assisted hoping. The extended and localized states are separated by a sharp boundary called mobility edge. A large change in carrier mobility occurs across this boundary. Several models have been proposed to explain the special features of glassy materials like chemical impurity independent electrical conductivity, pinned Fermi level in the middle of the band gap, and distribution of the density of states, similar to crystals. The concept of localized states in band tails and mobility edges is a common feature of these models. The Cohen-Fritzsche-Ovshinsky (CFO) proposed a model according to which compositional and topological disorders lead to extensive tailing. An appreciable density of states (DOS) in the middle of a band gap region is assumed to be due to the over lapping of conduction and valence band tails of the localized states (Fig. **3**). A DOS minimum lies in the middle of gap at which Fermi level lies.

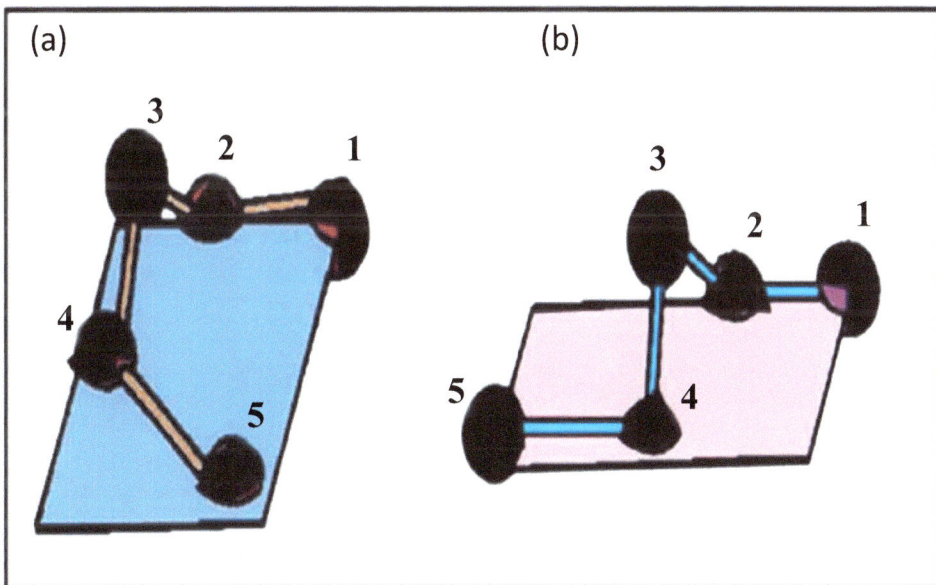

Fig. (1). Cis-coupling configuration (**a**) and Trans-coupling configuration (**b**) for molecular bonding in Selenium.

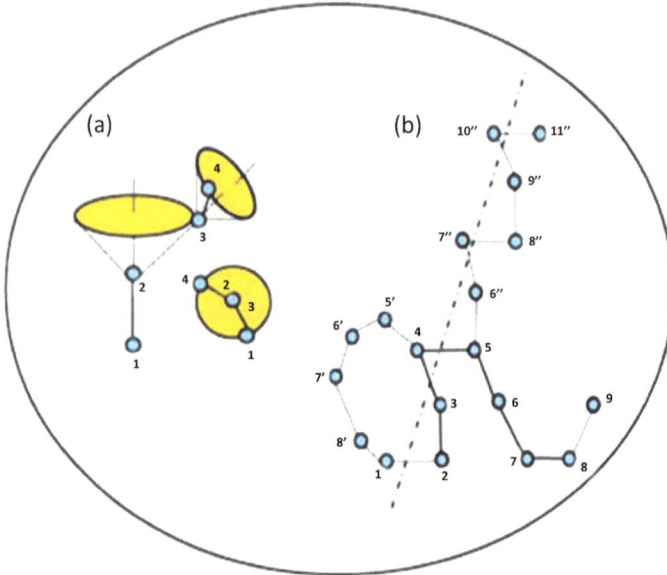

Fig. (2). **(a)** Free rotation chain model **(b)** Disordered chain model of Lucovsky and Galeener of selenium molecules [9].

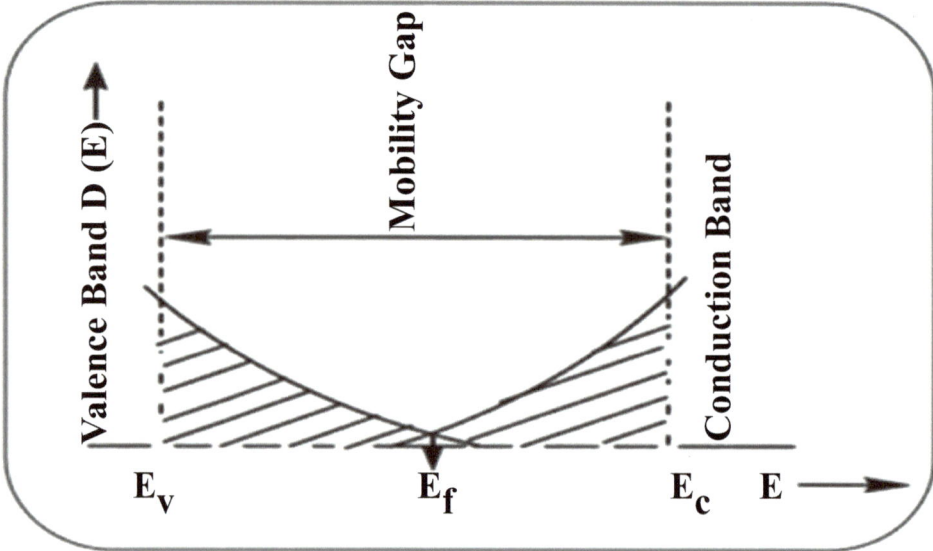

Fig. (3). CFO-Model of the density of states (DOS) in amorphous semiconductors.

Davis and Mott [20, 21] proposed that disorder induced states extend only a few tenths of an electron volt into the forbidden gap and Fermi level is pinned due to the existence of a band of compensated levels near the middle of the band gap (Fig. **4**); the origin of these levels is the defects in the random network.

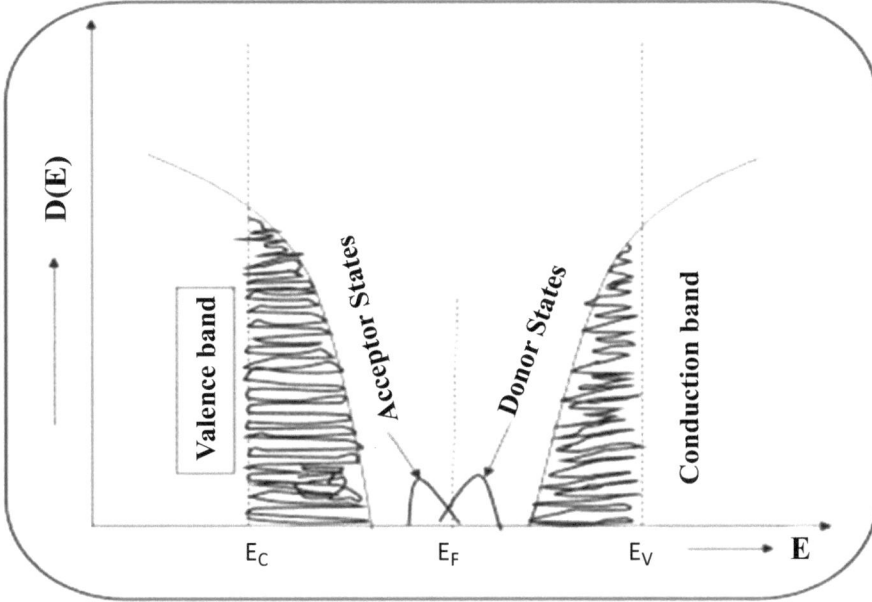

Fig. (4). Davis-Mott (DM) Model of density of states (DOS) in amorphous semiconductors.

Defects in Glasses

Overly connected or disconnected atomic bonds or a point defect-like structure is the defect in non-crystalline solids. Usually defects are rare (being smaller than ~1% of the total bonds) and hence can be ignored during the determination of structural properties, *e.g.* elastic constant and heat capacity at room temperature, however, defects may significantly affect electronic, optoelectronic properties if the states are produced in the band gap. In chalcogenides, like bonds in stoichiometric compositions [22], dangling bonds, are the well-known defects, and Raman scattering and electron spin resonance (ESR) have been used to detect the defects. The ESR can detect only unpaired electrons, for example D^0 (Neutral dangling bond) defects in chalcogenide glasses [23]. In case of pure Se and As-chalcogenides, ESR signal ($<10^{16}$ cm^{-3}) is absent in the dark, reflecting the fact that there are no neutral dangling bonds having unpaired electrons. However, ESR signal appears under illumination at low temperatures. Mott attributed to this to the presence of charged dangling bonds (D^+ and D^-) in the dark. Other indirect methods which have greater sensitivity are X-ray fluorescence, plasma emission and infrared absorption. The most important defects are charged defects. The following models explain the formation of charged defects in amorphous semiconductors.

Street-Mott Model

In the Street and Mott [24] model, dangling bond which normally contains an unpaired electron, may be occupied by zero, one or two electrons and these states are denoted as D^+, D^o, and D^-, respectively. Super scripts denote the total charge of the centre. It may be thought that D^o has an electron and one hole and hence neutral. D^+contains two holes (two empty orbitals). The defect centers are positively or negatively charged and neutral dangling bond can be achieved by excitation only because the following reaction is exothermic:

$$2D^0 = D^- + D^+$$

Mott and Street applied Holstein's model of diatomic molecules. Anderson found that the energy of the electron, localized at the site, is reduced due to the interaction with the lattice vibrations. Repulsive energy (columbic) of pairing electrons is also reduced. For sufficient strong coupling, effective potential energy (U_{eff}) becomes negative (electron pairing becomes favorable) and hence the above reaction is exothermic. So, this model provides more bonding states besides being empty and doubly occupied. Fig. (**5**) shows the transformation of two dangling bonds into two charged defects.

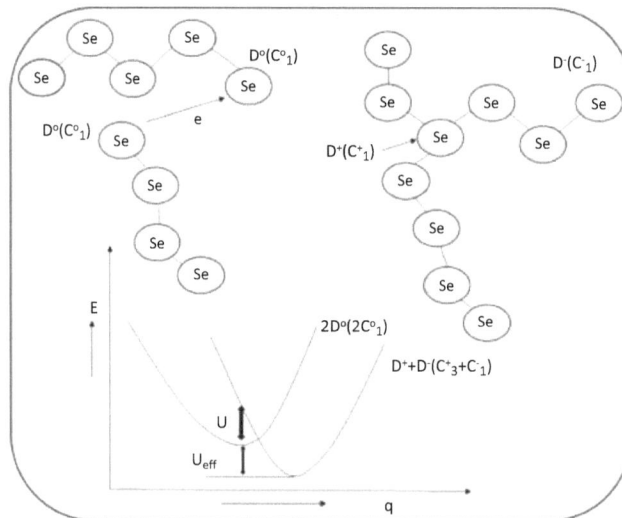

Fig. (5). Formation of charged dangling bonds from chain ends in a-Se.

Kastner, Adler and Fritzsche Model

This model assumes the formation of unusual bonding configurations, called valence alternation pairs [25], which are the result of specific interactions between nonbonding orbitals. Three-fold coordinated C_3^+ and singly coordinated C_1^- chalcogen atoms are the lowest energy defect states in chalcogenides. In the absence of other defects, these are equal in number. The formation of these defects can be represented as a reaction:

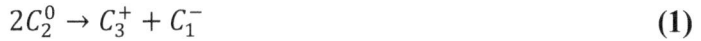

$$2C_2^0 \rightarrow C_3^+ + C_1^- \tag{1}$$

Since in the above reaction valency of the product chalcogen atom is changed from normal (2) to +1 and -1, these product defects are called valence alternation pairs (VAP). The density of VAP is given by:

$$[C_3^+][C_1^-] = N_A^2 \exp\left(- E_{VAP}/kT_e\right) \tag{2}$$

where E_{VAP} is the energy required for reaction (1)

Other important reactions of this model are:

(a) Release of an electron to the conduction band by C_3^0 :

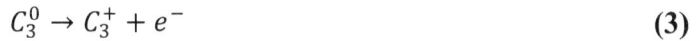

$$C_3^0 \rightarrow C_3^+ + e^- \tag{3}$$

(b) Inter conversion of defect which leads to gain in bonds:

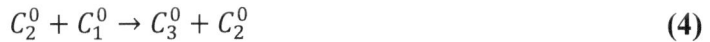

$$C_2^0 + C_1^0 \rightarrow C_3^0 + C_2^0 \tag{4}$$

(c) Creation of hole by C_3^0

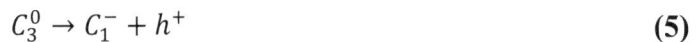

$$C_3^0 \rightarrow C_1^- + h^+ \tag{5}$$

(d) Loosing of coordination:

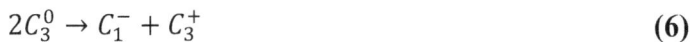

$$2C_3^0 \rightarrow C_1^- + C_3^+ \tag{6}$$

(e) Lifting of electron from C_1^- to conduction band:

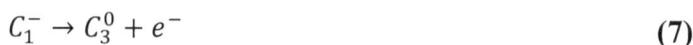

$$C_1^- \rightarrow C_3^0 + e^-$$ (7)

METHODS OF PREPARATION

A glass can be prepared by rapidly cooling the melt in a manner that avoids crystallization. Suitable methods are; melt quenched, thermal evaporation and microwave irradiation. A more commonly used method is the melt quenched technique. This is also known as sudden cooling of the melt. It is a fast process and restricts the nuclei to grow in a crystal-like minimum energy and entropy structure due to the unavailability of sufficient time for rearrangement. Glass formation becomes more feasible at the faster cooling rate, smaller volume of the sample and slower rate of crystallization. The glass formation tendency has been increased by chalcogenide compounds and alloys with predominantly covalent chemical bonds. The procedure of this technique includes weighing the desired amounts of alloying elements according to their suitable atomic weight percentages and then sealing them into cleaned quartz ampoules. The filled quartz ampoules have to be evacuated before sealing under a vacuum of 10^{-6}Torr to prevent any possible reaction of alloying elements with oxygen and other environmental impurities at a sufficient high temperature. Ampoules are heated by placing them into a furnace. The temperature of the furnace is increased in small steps above the melting point of alloying elements and kept constant for several hours. Ampoules have to be rocked frequently to make the melt homogeneous. After keeping the ampoules for the desired time in furnace, ampoules are quenched quickly by dropping the ampoules into ice-cooled water. The ingots of glassy materials are taken out by breaking the ampoules and crushed into fine powder.

PROPERTIES OF CHALCOGENIDE GLASSES

Thermal properties not only provide information regarding the structure of a glass but also determine the usefulness of a specific composition. The area of thermal properties includes studies related to glass transition, crystallization, melting and the stability to devitrification against thermal variation. Glass transition temperature and relaxation kinetics below glass transition temperature are two important characteristic features of glasses. It is observed that glass transition temperature depends on the cooling rate of the melt or on the heating rate when the glass is reheated in differential scanning calorimetry (DSC) [26]. Moynihan developed a method, which is being used extensively and uses the aspects of structural relaxation, and heating rate dependence of glass transition temperature (T_g), to estimate the activation energy of glass transition [9, 27]. Kissinger developed a method for the determination of activation energy of crystallization by considering the heating rate dependence of crystallization temperature [28] and commonly known as the peak shift method. An extensive literature exists which relates T_g to the magnitude of cohesive forces within the network because it is necessary to overcome these forces to initiate the atomic movement.

The differential scanning calorimetry (DSC) technique is the technique that is generally employed to study thermal properties. In this technique, few miligrams of a glassy sample, and an empty reference pan are heated at a constant heating rate, differential heat flow is recorded with respect to temperature, and during this heating, structural relaxations take place. Finally, crystallization takes place (exothermic peaks), and on further heating, it melts. DSC scan of a typical chalcogenide glass consists of an endothermic peak. Before crystallization, the temperature corresponding to this peak is called glass transition temperature (T_g). The comparative study of DSC scan of various samples or scan of a particular sample at different heating rates can be done with the help of parameters like the average coordination number, the overall mean bond energy, the heat of atomization, the band gap or the average group number or kinetics [11, 29]. T_g is directly related to the average coordination number by an empirical relationship: $\ln Tg \approx 1.6 <r> + 2.3$ [30]. Considering the effect of cross-linking in multi-component chalcogenide glasses, a modified Gibbs-Dimarzio equation is developed for predicting T_g at low average coordination numbers [14, 31]. Other physical properties which are related to glass transition temperature are mean atomic volume, Urbach tail width, melting temperature, nonlinear optical properties [16], *etc.* A higher glass transition temperature is predicated for a glassy sample with a higher mean bonding energy because a greater energy is required to relax the structure and break the bonds. Crystallization of a glass is also a kind of structural relaxation which is exothermic in terms of heat. Temperature corresponding to crystallization peak (T_c) is called crystallization temperature. The difference $\Delta T = T_c - T_g$ can be used to measure the resistance to devitrification of a glassy alloy or thermal stability of glass composition.

Chalcogenide glasses are well-known for their high transmittance in the near-infrared (NIR) and mid-infrared (MIR) regions and photo induced effects. Selenide and telluride glasses are opaque in the visible region, however, sulphide glasses are transparent in the long wavelength part of the visible region, up to which wavelength, a glass is transparent, it depends upon the chalcogen element that particular glass contains [32]. The refractive index (n), optical band gap (E_g^{opt}) and the absorption coefficient (α) are the important optical properties. The refractive index varies with the wavelength of the incident beam. The value of the refractive index for chalcogenide materials lies between 2 to 3 (low phonon energy materials) [33, 34]. Large refractive index makes these materials useful for photonics application. Rayleigh scattering losses are larger in these glasses in comparison to silica glasses. Fresnel reflection loss for air/glass interface varies between 10 to 25%. Optical absorption spectra of glassy materials generally consist of three regions, high absorption region ($\alpha \geq 10^4$ cm^{-1}) which is attributed to transition across the band gap, intermediate region (1 cm^{-1} $\leq \alpha \leq 10^4$ cm^{-1}) that is caused due to the defects and disorder, the third region is the weak absorption region ($\alpha \leq 1$ cm^{-1}). The intermediate absorption region is generally exponential in nature and expressed as:

$$\alpha = \alpha_0 \exp\left(\frac{h\nu}{E_U}\right) \tag{8}$$

Pre-exponential factor α_0 is a constant; E_U called the Urbach energy [35] and corresponds to the width of the tail of localized states in the energy gap. The absorption coefficient is related to the optical band gap by the Tauc relation [36]:

$$(\alpha h\nu)^{1/n} = B^{1/2}\left(h\nu - E_g^{opt}\right) \tag{9}$$

Where, E_g^{opt} is the optical band gap of the material, n is a constant, n= 2, 1/2 corresponds to the indirect allowed transition, and direct allowed transition. For most of the known chalcogenide glasses, n is found to be 2 [37]. The optical band gap can be obtained by plotting a graph between $(\alpha h\nu)^{1/n}$ *verses* photon energy (*h\nu*). In case of chalcogenide glasses, the value of electrical and optical band gap is the same due to the small interaction between electrons and holes.

APPLICATIONS OF CHALCOGENIDE GLASSES

Chalcogenide glasses (ChGs) have been studied as promising potential materials due to their exciting structural, thermal, optical and electronic properties [37, 38]. ChGs and ChGs based nano structured materials have several potential applications in industries to develop devices, medicals and security (communication) areas. In order to transmit the information between active elements through optical signals such as optical waveguides, optical fibres, these materials can be used to develop optical microcircuits by using photo darkening and photodoping methods [39]. The future optical computers, nonlinear elements (semiconductors), dye-sensitised solar cells (DSSCs), xerographic and thermoplastic media, photo-resistant [40], *etc.* can be produced at industrial level by using ChGs. The IR transparency of ChGs makes them suitable materials to develop a wide variety of optical applications. In ChGs materials, the refractive index in transparent wavelength regions increases (red shift), and follows the Kramers-Kronig relation. Structural properties, elastic properties and chemical properties can also be modified with illumination and their properties can be restored by annealing. By the illumination of light (hμ=Eg), the energy band gap of ChGs is assumed to be modified, and sub-gap light, with photon energy lying in the Urbach-tail region, also produces some changes and these changes are prominent than those induced by band gap light [41, 42]. Some optical phenomenon can also be recorded on ChGs based materials as a bulk and in the thin film form. Due to the photosensitive properties of these materials, they can be used to produce high-resolution imaging and photolithographic resistance. These materials are chemically stable in the air and can be used to develop a long core-clad optical fiber cable [43]. ChGs have excellent properties, and DSSCs based on these materials have overall excellent performance with commercial viability. In the future, these materials can be used for the fabrication of solar cells in place of silicon (Si).

Worldwide, due to the high consumption of energy obtained from natural sources and gradually increasing environmental problems, the research on energy storage in the form of green energy (electrochemical energy) increases tremendously. In the last two decades, the technology of thin-film formation has drawn much attention, and second-generation solar cells, having QDs based sensitizing materials began to flourish [44 - 47]. Several materials can be deposited easily by using thin-film deposition technique on any substrate like metals, polymers and glassy materials. Inorganic compounds like chalcogenide glasses offer outstanding potential applications compared to organic materials due to their unique electronic, optical and thermal properties. Dye-sensitized solar cell (DSSC) is a low-cost solar cell that converts light into electricity based on wide band gap semiconductor sensitisation and belongs to a group of thin film solar cell [47]. DSSC consists of four primary parts, *i.e.*, working electrode, dye (or sensitizer), electrolyte and counter electrode. The typical construction of DSSC involves a working electrode soaked with a sensitizer (or a dye) and sealed to the counter electrode with a thin film of electrolyte placed between the working and counter electrode (Fig. **6**).

The sensitizer is one of the major components of the DSSCs. Chalcogenide based semiconducting quantum dots (QDs) are the materials with promising use as sensitizers due to their high optical absorption coefficients, tunable energy bandgaps, multiple exciton generation properties and large dipole moments [46, 47]. The most commonly used QD sensitizers can be classified into three categories: (i) Cadmium-ChGs based QD sensitizers including cadmium-sulfur (Cd-S), cadmium-selenium (Cd-Se), cadmium tellurium (Cd-Te) and their nano-crystal alloys [46 - 48]; (ii) Lead-ChGs based Q Desensitizers: In this category lead based materials can be used like lead-sulfur (Pb-S), led-selenium (Pb-Se); and (iii) Antimony-ChGs based QD sensitizers such as antimony-sulfide, Sb_2S_3 QDs [49 - 56].

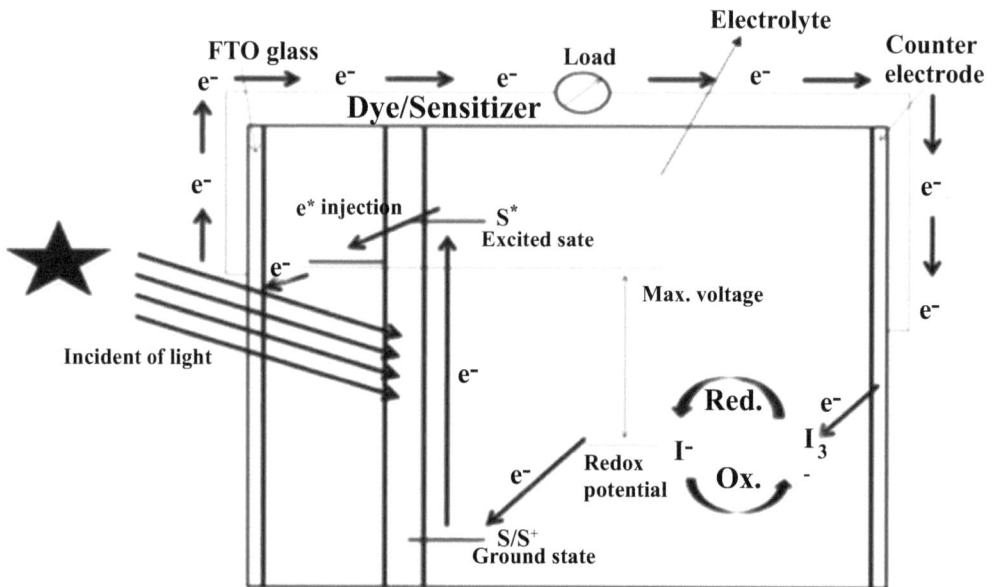

Fig. (6). Schematic diagram of dye-sensitized solar cell (DSSC).

The properties of chalcogenide materials can be tuned to satisfy the requirements of optical data storage devices like PC disk. For an optical data storage system, the properties of ChGs such as the transformation of phase from amorphous to crystalline or crystalline to amorphous, large signal to noise ratio between these phases, stability of amorphous phase and the re-cyclability play a significant role. Recyclability means the ability to write the information million times on the same track without any degradation in the optical response of the PC disk. In chalcogenide materials, the crystallization behavior/speed strongly depends on the exact composition ratio of the materials such as in GeTe binary system, the atomic mole percent ratio 50:50 is used *i.e.* $Ge_{50}Te_{50}$ [54, 57]. For example, in GeTe binary system, the crystallization speed depends on the atomic ratio between Ge and Te and it has a wide range (nano second to milli second). Ovshinsky *et al.* showed that the addition of antimony, Sb in binary GeTe system resolves the condition of exact atomic ratio and makes it possible to fabricate PC material for data storage like Ge–Sb–Te (known as GST materials having three stoichiometric phases $Ge_1Sb_4Te_7$, $Ge_1Sb_2Te_4$ and $Ge_2Sb_2Te_5$) [58, 59]. For optical data storage system, the selection of material has the desired crystallization speed and crystallization temperature because these two parameters are correlated to each other (when crystallization speed is high, the crystallization temperature is lowered). In ChG based optical storage system, the written mark is unstable due to the amorphous state (meta stable state) of the materials having low crystallization temperature while for those materials having high crystallization temperature *i.e.* longer crystallization time. These two behaviors of such materials make them unattractive for data storage purposes.

The properties of amorphous solids depend on the nature of the chemical bonding between the adjacent atoms and their electronic configuration. Due to these properties, the amorphous materials behave like insulating and or semiconducting materials (Fig. 7). In these materials, the wave-functions of electrons and holes at band edges and mid-gaps are localized. Therefore, the electron (hole) mobility in amorphous material (non-crystalline) becomes smaller than that in the corresponding crystal, because the localized electrons (hole) in the band edges govern the mobility [56 - 59]. Many amorphous materials are found to exhibit significantly low electrical conduction and are commonly known as insulators having energy gaps of 5-10 eV. On the other hand, amorphous ChGs and tetrahedral materials such as a-Si: H have the energy gap of 1-3 eV and are known as semiconductors.

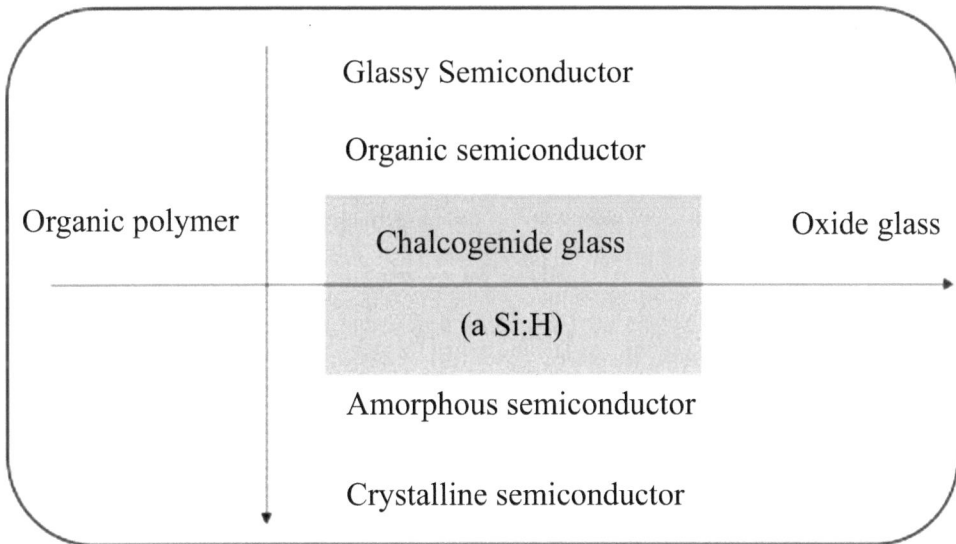

Glassy Semiconductor

Organic semiconductor

Organic polymer   Chalcogenide glass   Oxide glass

(a Si:H)

Amorphous semiconductor

Crystalline semiconductor

Fig. (7). Chalcogenides as a glass (horizontal) and a semiconductor (vertical).

Amorphous semiconductors are divided into two groups; Ionic bonded semiconductors like halide and oxide glasses; covalently bonded amorphous semiconductors. Covalently bonded amorphous semiconductors are further divided into the following categories: (i) Terahedrally bonded amorphous semiconductors like amorphous Silicon, Germanium (ii) Chalcogenide amorphous semiconductors that contain one or more chalcogen elements as a major constituent of alloy.

Chalcogenide glasses (ChGs) have drawn much attention in recent years due to their attractive optical, thermal, electronic and electrical properties such as transmission in middle and far infrared regions, lower values of phonon energies, higher values of refractive indices, high values of optical non linearity, laser induced crystallization, and memory switching [1 - 6]. A Physicist B.G. Kolomiets and a Chemist N.A. Gorjunova [60] from Ioffe's Physical and Technical Institute (St. Petersburg) have discovered the semiconducting properties of chalcogenide glasses during the survey of photoconductivity of these materials. At the same time, researchers at Leningrad state university, studied the chalcogenide glass from chemical point of view, *i.e.*, as ion-conducting [61] and infrared transmitting materials [26]. Chalcogenide alloys can be regarded as a soft semiconductor, 'soft' because they have a flexible and viscous (due to two-fold coordination number) atomic structure and 'semiconductor' because they possess an energy band gap (1~2 eV) like semiconductor materials (1~3 eV) [62] and hence can transmit certain frequencies of light spectrum (mostly infrared). Generally, these amorphous materials are p-type semiconductors [63]. The reason for the same is considered to be two-fold; (i) the carrier life time excited above the Fermi level (electrons) is small as compared to below the Fermi level (holes) as determined with the photoconductivity experiment, (ii) Seeback coefficient is positive. The disorder-induced localized states extended into the forbidden band gap have a significant effect on the electrical and optical properties of these glassy materials.

Phase-change memory consists of chalcogenide alloys that change from a disordered, amorphous structure to a crystalline structure when an electrical pulse is applied. This switching occurs rapidly at a certain threshold voltage, known as threshold switching [63]. The material has high electrical resistance in its amorphous state and low resistance in its crystalline state — corresponding to the 1 and 0 states of binary data. The chalcogenides that are mostly investigated as phase change materials are Ge-Sb-Te ternary alloys [1, 55, 64 - 66]. They exist in three phases, one amorphous and two crystalline namely face-centred cubic (fcc) and hexagonal closed packing (hcp). The amorphous to fcc transition occurs at 120°- 180°C, this is a meta-stable state having high resistivity and low reflectivity. Phase-change memory is being actively pursued as an alternative to ubiquitous flash memory for data storage applications because flash memory is limited in its storage density when devices get smaller than 20 nanometers. On the other hand, a phase-change memory device can be less than 10 nanometers, allowing more memory to be squeezed into tinier spaces. Data can be written into phase-change memories very quickly.

CONCLUSION

Chalcogenide glasses (ChGs) have unique structural, thermal, optical and electronic properties and the functionalities of these materials will revolutionize many research areas and industrial applications. ChGs are transparent in nature from the visible to the near infrared region and can be used for the preparation of optical and electronic devices like CG fibers, optical switches, sensors and phase charge memorizers. For the preparation of ChGs, several techniques were adopted in which melt quench technique is one of the most effective techniques. Due to the presence of localized states associated with structural properties like energy band gap and the presence of different defects, ChGs have photoinduced optical properties like photoluminescence and photo darkening. The applications of CGs especially in semiconductors, electrical memories, phase change materials and dye sensitized solar cell (DSSCs) are discussed.

CONSENT FOR PUBLICATION

Not applicable.

CONFLICT OF INTEREST

The authors declare no conflict of interest, financial or otherwise.

ACKNOWLEDGEMENTS

The authors HK and ALS are thankful to DST-FIST and IoE scheme No. 631 for providing financial assistance through PURSE and IoE grants.

REFERENCES

[1]   Borisova ZU. Glassy semiconductors. Plenum New York 1981.
        [http://dx.doi.org/10.1007/978-1-4757-0851-6]

[2]   Shelby JE. Introduction to Glass Science and Technology. Chambridge, U. K: Royal Society of Chemistry 2005.

[3]   Seddon AB. Chalcogenide glasses: a review of their preparation, properties and applications. J Non-Cryst Solids 1995; 184: 44-50.
        [http://dx.doi.org/10.1016/0022-3093(94)00686-5]

[4]   Andriesh AM, Iovu MS, Shutov SD. Chalcogenide non-crystalline semiconductor in opto-electronics. J Optoelectron Adv Mater 2002; 4: 631-47.

[5]   Mehta N. Application of chalcogenide glasses in electronics & optoelectronics: A review. J Sci Ind Res (India) 2005; 65: 777-86.

[6]   Prakash T. Review on nanostructured semiconductors for dye sensitized solar cells. Electron Mater Lett 2012; 8(3): 231-43.
[http://dx.doi.org/10.1007/s13391-012-1038-x]

[7]   Ovshinsky SR, Adler D. Local structure, bonding, and electronic properties of covalent amorphous semiconductors. Contemp Phys 1978; 19(2): 109-26.
[http://dx.doi.org/10.1080/00107517808210876]

[8]   Aivasov AA, Budogyn BG, Vikhrov SP, Popov AI. Disorder semiconductors (Neuporyadochenneepoluprovodniki). Moscow: Vesshayashkola 1995.

[9]   Lucovsky G, Galeener FL. Intermediate range order in amorphous solids. J Non-Cryst Solids 1980; 35-36: 1209-14.
[http://dx.doi.org/10.1016/0022-3093(80)90362-2]

[10]   Young PA, Thege WG. Structure of evaporated films of arsenic trisulphide. Thin Solid Films 1971; 7(1): 41-9.
[http://dx.doi.org/10.1016/0040-6090(71)90012-5]

[11]   Tominaga J, Haratani S, Handa T, Yanagiuchi K. Scanning tunneling microscopy image of $GeSb_2Te_4$ thin films. Jpn J Appl Phys 1992; 31(Part 2, No. 6B): L799-802.
[http://dx.doi.org/10.1143/JJAP.31.L799]

[12]   Takebe H, Maeda H, Morinaga K. Compositional variation in the structure of Ge–S glasses. J Non-Cryst Solids 2001; 291(1-2): 14-24.
[http://dx.doi.org/10.1016/S0022-3093(01)00820-1]

[13]   Kaplow R, Rowe TA, Averbach BL. Atomic arrangement in vitreous selenium. Phys Rev 1968; 168(3): 1068-79.
[http://dx.doi.org/10.1103/PhysRev.168.1068]

[14]   Cherkasov Yu. A. and Kreitor, L.G. Exiton absorption and photoconductivity in amorphous analogs of various crystalline forms of selenium. Fizika Tverdogo Tela 1974; 16: 2407-10.

[15]   Richter H, Breiting G. Verschiedene Formen von amorphen Selen. Z Naturwiss Med Grundlagenforsch 1971; 26a: 1699-708.

[16]   Malaurent JC, Dixmier J. A Random Chain Model for Amorphous Selenium, Thestructure of noncrystalline materials. Cambridge, UK: Cambridge University Press 1977; pp. 49-51.

[17]   Popov A. 2004.

[18]   Mott NF, Davis EA. Electronic Processes in Non- Crystalline Materials. Oxford: Clarendon 1979.

[19]   Davis EA, Mott NF. Conduction in non-crystalline systems V. Conductivity, optical absorption and photoconductivity in amorphous semiconductors. Philos Mag 1970; 22(179): 0903-22.
[http://dx.doi.org/10.1080/14786437008221061]

[20]   Elliott SR. Physics of Amorphous Materials. 2nd ed., Essex: Longman Scientific & Technical 1990.

[21]   Phillips JC. Topology of covalent non-crystalline solids I: Short-range order in chalcogenide alloys. J Non-Cryst Solids 1979; 34(2): 153-81.
[http://dx.doi.org/10.1016/0022-3093(79)90033-4]

[22]   Fritzsche H. Optical and electrical energy gaps in amorphous semiconductors. J Non-Cryst Solids 1971; 6(1): 49-71.
[http://dx.doi.org/10.1016/0022-3093(71)90015-9]

[23]   Halpern V. Localized electron states in the arsenic chalcogenides. Philos Mag 1976; 34(3): 331-5.
[http://dx.doi.org/10.1080/14786437608222026]

[24]   Arai K, Hattori Y, Namikawa H, Saito S. Effects of densification on electrical and optical properties of chalcogenide glasses. Jpn J Appl Phys 1973; 12(11): 1717-22.
[http://dx.doi.org/10.1143/JJAP.12.1717]

[25]   Street RA, Mott NF. States in the Gap in Glassy Semiconductors. Phys Rev Lett 1975; 35(19): 1293-6.
       [http://dx.doi.org/10.1103/PhysRevLett.35.1293]

[26]   Kastner M, Adler D, Fritzsche H. Valence alteration model for localized gap states in loan pair semiconductors. Phys Rev Lett 1976; 37(22): 1504-7.
       [http://dx.doi.org/10.1103/PhysRevLett.37.1504]

[27]   Ovshinsky SR, Adler D. Local structure, bonding, and electronic properties of covalent amorphous semiconductors. Contemp Phys 1978; 19(2): 109-26.
       [http://dx.doi.org/10.1080/00107517808210876]

[28]   Aivasov AA, Budogyn BG, Vikhrov SP, Popov AI. Disorded semiconductors (Neuporyadochenneepoluprovodniki). Moscow: Vesshayashkola 1995.

[29]   Young PA, Thege WG. Structure of evaporated films of arsenic trisulphide. Thin Solid Films 1971; 7(1): 41-9.
       [http://dx.doi.org/10.1016/0040-6090(71)90012-5]

[30]   Takebe H, Maeda H, Morinaga K. Compositional variation in the structure of Ge–S glasses. J Non-Cryst Solids 2001; 291(1-2): 14-24.
       [http://dx.doi.org/10.1016/S0022-3093(01)00820-1]

[31]   Kaplow R, Rowe TA, Averbach BL. Atomic arrangement in vitreous selenium. Phys Rev 1968; 168(3): 1068-79.
       [http://dx.doi.org/10.1103/PhysRev.168.1068]

[32]   Richter H, Breiting G. VerschiedeneFormen von amorphenSelen. Z Naturwiss Med Grundlagenforsch 1971; 26a: 1699-708.

[33]   Savage JA. Infrared Optical Materials and their Antireflection Coatings. Bristol: Adam Hilger 1985.

[34]   Kadono K. Nonoxide Glass-forming systems-glass formation and structure, and optical properties of rare-earth ions in glasses. J Ceram Soc Jpn 2007; 115(1341): 297-303.
       [http://dx.doi.org/10.2109/jcersj.115.297]

[35]   Urbach F. The long-wavelength edge of photographic sensitivity and of the electronic absorption of solids. Phys Rev 1953; 92(5): 1324.
       [http://dx.doi.org/10.1103/PhysRev.92.1324]

[36]   Tauc J. Amorphous and liquid semiconductors. New York: Plenum Press 1974.
       [http://dx.doi.org/10.1007/978-1-4615-8705-7]

[37]   Aly KA, Abd Elnaeim AM, Afify N, Abousehly AM. Improvement of the electrical properties of Se_3Te_1 thin films by In additions. J Non-Cryst Solids 2012; 358(20): 2759-63.
       [http://dx.doi.org/10.1016/j.jnoncrysol.2012.06.029]

[38]   Fairman R, Ushkov B. Semiconducting Chalcogenide Glass 1. Netherlands: Elsevier Academic Press 2004.

[39]   Hilton AR. Chalcogenide Glasses for Infrared Optics, McGrawHill Companies, Inc. 2010.

[40]   Mehta N. Applications of chalcogenide glasses in electronics and optoelectronics: A review. J Sci Ind Res (India) 2006; 65: 777-86.

[41]   Seddon AB. Chalcogenide glasses: a review of their preparation, properties and applications. J Non-Cryst Solids 1995; 184: 44-50.
       [http://dx.doi.org/10.1016/0022-3093(94)00686-5]

[42]   Singh PK, Dwivedi DK. Chalcogenide glass: Fabrication techniques, properties and applications. Ferroelectrics 2017; 520(1): 256-73.
       [http://dx.doi.org/10.1080/00150193.2017.1412187]

[43]   Fritzsche H. Light induced effects in glasses.Insulating and Semiconducting Glasses. Singapore: World Scientific 1993.

[44]   Shimakawa K, Kolobov A, Elliott SR. Photoinduced effects and metastability in amorphous semiconductors and insulators. Adv Phys 1995; 44(6): 475-588.
[http://dx.doi.org/10.1080/00018739500101576]

[45]   Sanghera JS, Florea CM, Shaw LB, *et al.* Non-linear properties of chalcogenide glasses and fibers. J Non-Cryst Solids 2008; 354(2-9): 462-7.
[http://dx.doi.org/10.1016/j.jnoncrysol.2007.06.104]

[46]   Ye M, Wen X, Wang M, *et al.* Recent advances in dye-sensitized solar cells: from photoanodes, sensitizers and electrolytes to counter electrodes. Mater Today 2015; 18(3): 155-62.
[http://dx.doi.org/10.1016/j.mattod.2014.09.001]

[47]   Jun HK, Careem MA, Arof AK. Quantum dot-sensitized solar cells—perspective and recent developments: A review of Cd chalcogenide quantum dots as sensitizers. Renew Sustain Energy Rev 2013; 22: 148-67.
[http://dx.doi.org/10.1016/j.rser.2013.01.030]

[48]   Jacoby M. The future of low-cost solar cells. Chem Eng News 2016; 94: 30-5.

[49]   Rhee JH, Chung CC, Diau EWG. A perspective of mesoscopic solar cells based on metal chalcogenide quantum dots and organometal-halide perovskites. NPG Asia Mater 2013; 5(10): e68.
[http://dx.doi.org/10.1038/am.2013.53]

[50]   Diguna LJ, Shen Q, Kobayashi J, Toyoda T. High efficiency of CdSe quantum-dot-sensitized TiO_2 inverse opal solar cells. Appl Phys Lett 2007; 91(2): 023116.
[http://dx.doi.org/10.1063/1.2757130]

[51]   Chen Z, Augustyn V, Wen J, *et al.* High-performance supercapacitors based on intertwined CNT/V_2O_5 nanowire nanocomposites. Adv Mater 2011; 23(6): 791-5.
[http://dx.doi.org/10.1002/adma.201003658] [PMID: 21287644]

[52]   Lee YL, Lo YS. Highly efficient quantum-dot-sensitized solar cell based on co-sensitization of CdS/CdSe. Adv Funct Mater 2009; 19(4): 604-9.
[http://dx.doi.org/10.1002/adfm.200800940]

[53]   Ye M, Wen X, Wang M, *et al.* Recent advances in dye-sensitized solar cells: from photoanodes, sensitizers and electrolytes to counter electrodes. Mater Today 2015; 18(3): 155-62.
[http://dx.doi.org/10.1016/j.mattod.2014.09.001]

[54]   Kolobov AV, Fons P, Tominaga J, Ovshinsky SR. Vacancy-mediated three-center four-electron bonds in $GeTe-Sb_2Te_3$ phase-change memory alloys. Phys Rev B Condens Matter Mater Phys 2013; 87(16): 165206.
[http://dx.doi.org/10.1103/PhysRevB.87.165206]

[55]   Hegedüs J, Elliott SR. Microscopic origin of the fast crystallization ability of Ge–Sb–Te phase-change memory materials. Nat Mater 2008; 7(5): 399-405.
[http://dx.doi.org/10.1038/nmat2157] [PMID: 18362909]

[56]   Quenez P, Gorochov O. Préparation et propriétés de monocristaux de Cd_4GeSe_6. J Cryst Growth 1974; 26(1): 55-8.
[http://dx.doi.org/10.1016/0022-0248(74)90199-7]

[57]   Kuhs WF, Nitsche R, Scheunemann K. The argyrodites — A new family of tetrahedrally close-packed structures. Mater Res Bull 1979; 14(2): 241-8.
[http://dx.doi.org/10.1016/0025-5408(79)90125-9]

[58]   Chopra KL, Major S, Pandya DK. Transparent conductors—A status review. Thin Solid Films 1983; 102(1): 1-46.
[http://dx.doi.org/10.1016/0040-6090(83)90256-0]

[59]   Tanaka K. Ion conducting chalcogenide glasses: fundamentals and photo induced phenomena. Izv Him 1993; 26: 450-9.

[60] Kolomietsand BT, Gorjunova NA. The semiconducting properties of chalcogenide glasses. Soviet Journal of Technical Physics 1955; 25: 984-1015.

[61] Bletskan DI. Glass formation in binary and ternary chalcogenide systems. Chalcogenide Lett 2006; 3(11): 81-119.

[62] Mott, NF and Davis, EA, Electron Processes in Non-Crystalline Materials. Oxford: Clarendon Press 1979.

[63] Kolobov AV. On the origin of p-type conductivity in amorphous chalcogenides. J Non-Cryst Solids 1996. 198–200, 728–731.

[64] Elliott SR. Chalcogenide Phase-Change Materials: Past and Future. Int J Appl Glass Sci 2015; 1-4.

[65] Wuttig M, Yamada N. Phase-change materials for rewriteable data storage. Nat Mater 2007; 6(11): 824-32.
[http://dx.doi.org/10.1038/nmat2009] [PMID: 17972937]

[66] Chua EK, Shi LP, Zhao R, *et al.* Low resistance, high dynamic range reconfigurable phase change switch for radio frequency applications. Appl Phys Lett 2010; 97(18): 183506.
[http://dx.doi.org/10.1063/1.3508954]

CHAPTER 3

Al Doped ZnO Thin Films: Beginning to Developments Afoot

Kamakhya Prakash Misra[1,*]

[1] *Department of Physics, School of Basic Sciences, Manipal University Jaipur, Jaipur, Rajasthan-303007, India*

Abstract: In the last three decades, Zinc oxide (ZnO) has been found to be one of the most resourceful materials having tremendous potential applications in manifolds covering a wide variety of areas. It is continuously explored in different forms and structures. ZnO-based layers have an established place in the industry that ranges from protecting degradable items to detecting toxic gases. A wide variety of ZnO-based advanced coatings and their surface treatments along with innovative functionalization technologies offer a multitude of options for making them useful in diverse industries. Multiple techniques ranging from exceedingly sophisticated ones like molecular beam epitaxy and atomic layer deposition to highly-cost effective ones like sol-gel spin coating and dip coating, *etc.* have been used for developing the ZnO based thin films. Doping suitable elements into ZnO matrix is the most promising strategy to alter its properties drastically. Out of numerous dopants, Aluminum (Al) offers some of the excellent and reproducible features in ZnO films which make Al doped ZnO (AZO) a reputable system in industries like thin film transistor manufacturing and solar cells. Specifically, its established and repeatable behavior in terms of transparency and conductivity becauseis finding huge applications as a transparent conducting oxide (TCO). Extensive research on AZO coatings derived from different methods day-b--day opens up a new gateway for interesting perspectives by optimizing surface nanostructures. Here a brief account of historical developments of ZnO to AZO films along with their applications in certain key areas like TCOs, solar cells, thin film transistors, flexible electronics and plasmonics, *etc.* is presented.

Keywords: ZnO, TCO, AZO, TFTs, Plasmonics, Flexible Electronics.

INTRODUCTION

The ancient Indian medical text Charaka Samhita, dates of composition of which are uncertain, refers to the use of *pushpanjan*, most probably zinc oxide, as an

* **Corresponding author Kamakhya Prakash Misra:** Department of Physics, School of Basic Sciences, Manipal University Jaipur, Jaipur, Rajasthan-303007, India;
E-mails: misra.kamakhya@gmail.com, kamakhyaprakash.misra@jaipur.manipal.edu

Arti Srivastava, Mridula Tripathi, Kalpana Awasthi and Subhash Banerjee (Eds.)

ointment for eyes and open wounds. A book namely '2000 years of zinc and brass' by P. T. Craddock [1], which is supposed to be the first comprehensive review of the technical history of the production of zinc and brass, describes the inception and technical development of the processes by which zinc and brass were made at various centers around the world. Earlier zinc and zinc oxide (ZnO) were recognized and produced in India using a primitive form of the direct synthesis process. The main usage of zinc oxide also known as zinc white was in paints and as an additive to various ointments. Nowadays, a significant amount of ZnO (50% to 60% of its total applications) is finding numerous applications in the rubber industry [2]. Rest is being used in the ceramic industry [2], concrete manufacturing [3], skin ointments and sunblock creams, food (as a source of zinc), oil and pigments. An illustration shown in Fig. (**1**) depicts the applications of ZnO in several industries.

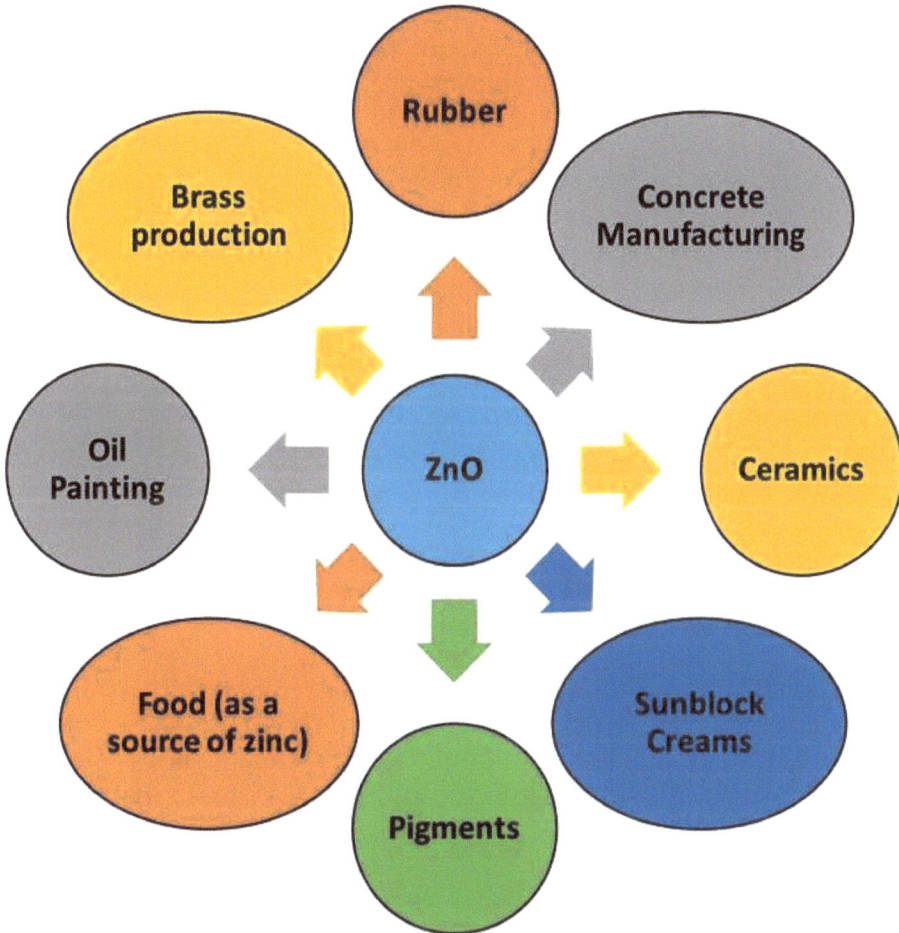

Fig. (1). Schematic illustration of use of ZnO in various industries.

A small portion of its total use is related to the study of its functional properties. Besides large-scale industries, it has been observed to be a promising candidate for thin film devices in the areas of optoelectronics [4, 5], piezoelectronics [6], transparent electronics [7], spintronics [8], sensing [9], and photovoltaics [10]. This is all possible due to its inherent characteristics and a wide range of achievable morphologies. The emergence of materials research has led to the modification of conventional compounds by alloying or doping. Currently, the traditional boundaries are being broken in forming solid solutions. ZnO has shown its potential to offer its matrix as a host, to a number of elements of the periodic table, which has opened tremendous opportunities for its applications as per the current needs. Despite greater achievements, two problems in the ZnO related technology are yet persistent: (1) the inability to achieve significant p-type conductivity which is reproducible and stable in time and (2) the development of almost incomprehensible diverse nanostructures with different properties under individual synthesis environment. Apart from these problems, the n-type conductivity of ZnO has a disputable origin, and p-type conductivity, if it is achieved, does not have a well-defined origin.

Several excellent features like higher transparency, tunable band gap and elevated conductivity *etc.* can be achieved by introducing a suitable dopant in the ZnO system or just from a non-stoichiometric configuration. Many metals such as Al, In, Ga, Ca, Mg, Sr and Ag have been found as an encouraging substitute of Zn in ZnO to achieve favorable properties. In the last two decades, a trivalent cation, Al has been highly popular as a dopant in ZnO because of the cheaper cost and ease of availability. Al-doped ZnO (AZO) has shown good thermal and chemical stability even under reducing ambient. It has emerged as a replacement of Indium Tin Oxide (ITO) on the commercial scale. Here, the importance and applications of AZO in various scientific fields along with the basic properties of ZnO are presented.

BASIC PROPERTIES OF ZnO

ZnO is probably the richest family of nanostructures among all one-dimensional nanostructures, including carbon nanotubes. In stoichiometric form ZnO, a wide-bandgap semiconductor belonging to the II-VI semiconductor group, is an insulator. Deviations from stoichiometry are, however, more the rule than the exception and this imparts semiconducting properties to metal oxides. As a best example, the oxygen vacancies or zinc interstitials in ZnO are found to make it an n-type semiconductor. A semiconductor material with a sufficiently wide band gap at room temperature makes it a candidate for optical sources in the visible region or UV region. It has several desirable properties like high transparency, high electron mobility [11], wide band gap (3.37 eV) [12 - 17], high exciton binding energy (60 meV) at room temperature [18] and strong room temperature luminescence [19]. It is an important functional oxide, exhibiting non-central symmetry, giving rise to piezoelectric behavior, which is a key property in building electromechanical coupled sensors and transducers. Besides, it is biocompatible, non-toxic, chemically and mechanically stable and available in abundance. Other than its wide usage in bulk form, thin films are used in many of the emerging applications, like in transparent conducting electrodes, liquid crystal displays, energy saving or heat protecting windows, thin film transistors and transparent photonics [19 - 24]. The diversity of nanostructures is day-by-day opening up many fields of research in nanotechnology for ZnO. To have an easier and comprehensive understanding, various properties of ZnO are illustrated in Fig. (**2**).

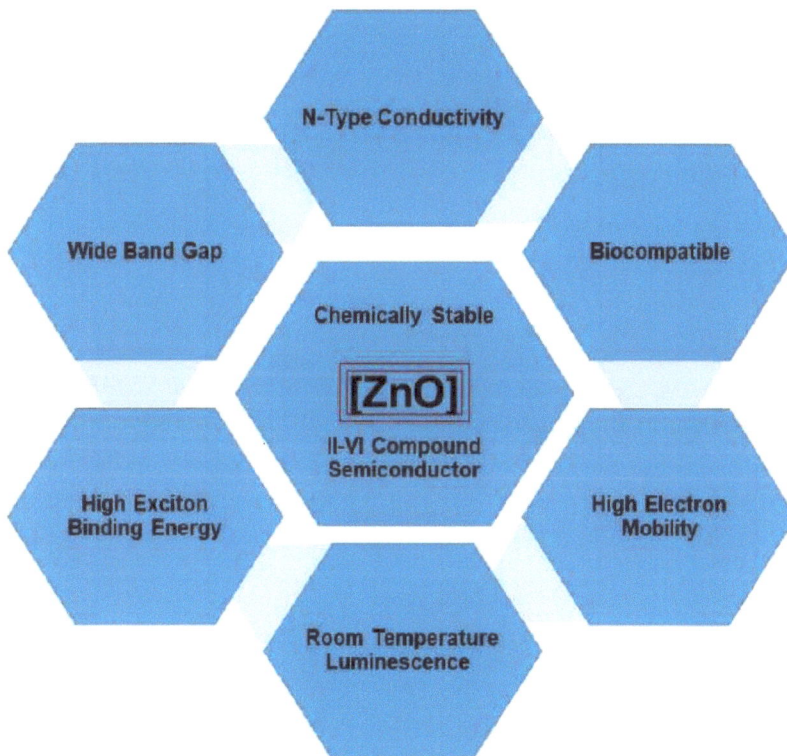

Fig. (2). Physical and chemical properties of ZnO.

Thermodynamically stable crystal structure of ZnO is hexagonal wurtzite belonging to the space group P63mc [25]. Looking at the structure, the lattice of ZnO has two intertwined hexagonal close-packed Zn and O lattices. The lattice arrangement is such that each Zn^{2+} ion is coordinated by four O^{2-} ions in a tetrahedral structure and vice-versa. The wurtzite structure is shown in Fig. (**3a**). In fact, such type of atomic arrangement results in a non-centrosymmetric ZnO crystal, which is why piezoelectric behavior in it, is noticed. The hexagonal prism showing various crystallographic faces is given (Fig. **4**). In addition to the wurtzite phase, ZnO is also seen to crystallize in the cubic zinc blende and rocksalt (NaCl) structures as shown in Fig. (**3b**) and (**c**). The sp^3 hybridized bonding is available in ZnO with equal ionic and covalent character.

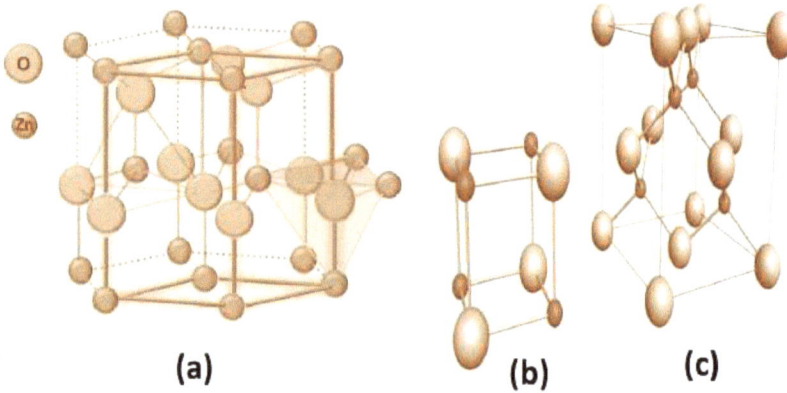

Fig. (3). (a) Hexagonal wurtzite (b) rock salt and (c) zinc blende structure of ZnO.

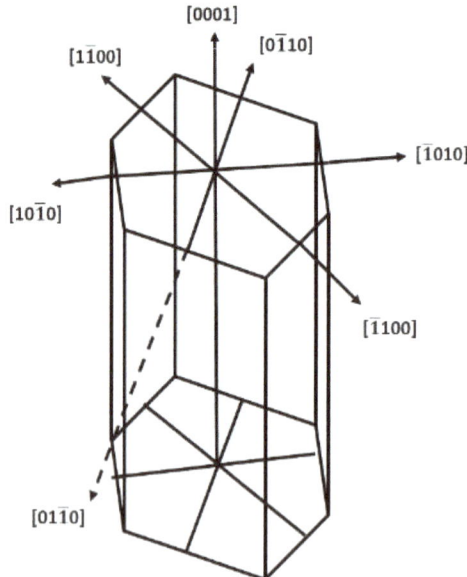

Fig. (4). Hexagonal prism demonstrating various crystallographic planes in ZnO.

Al-DOPED ZnO

Al-doped ZnO is commercially successful as a transparent conducting oxide (TCO) for transparent electronic applications [26]. Current research is targeting to optimize the electrical and optical characteristics of AZO films. The epitaxial and superlattice design resulting from AZO based films for any device structure has given an evidence of reduced losses at the layer interfaces eventually leading to improved performance of the device. The conductivity of ZnO can be greatly enhanced by doping with Al, B, Ga, or other impurities. But aluminum doped ZnO (AZO) being composed of aluminum (Al) and zinc (Zn), two common and inexpensive materials, has received great attention. AZO has demonstrated attractive applications in solar cells as TCO and rear contacts, owing to its low deposition temperature as well as its high stability. Polycrystalline AZO films with a columnar grain structure show no strong out-of-plane texture and random in-plane orientation. The texture has been found to be affected by the film-growth methods, their processes and/or deposition steps such as a multi-deposition process using a buffer layer and post-annealing. Polycrystalline AZO films are easily produced by the deposition methods such as magnetron sputtering [27 - 33], chemical vapour deposition [34], pulsed laser deposition [35 - 37], and the sol-gel method [38 - 41]. Nomoto *et al.* have shown that the grain boundaries play an important role in doping, as the foreign atoms change the number of active Al donors and free carriers in crystallites by acting as sites for dopant segregation and carrier trapping [42]. The carrier transport and the distribution of crystallographic orientations of polycrystalline AZO films [42 - 46] are the two factors of great importance. Two reasons are ascribed to this cause; one is the material's behaviour which hinders the free diffusion of the carrier and the other is practical technology to gain the requisite mobility of the carriers making AZO a suitable candidate for applications in electrical and optoelectronic devices [47 - 49]. Further, carrier diffusion has been found to be limited by two factors: the intrinsic carrier mobility in intra-grains and the contribution of grain boundary scattering to the carrier transport governed by the crystallographic orientation distribution [42 - 46].

Nowadays, the applications of AZO-films are widely extending to multiple areas of research. The illustration given in Fig. (5) presents a thorough overview of the applications of AZO films in various fields of science and technology. Here, three major areas of applications of AZO are elaborately discussed.

Fig. (5). Various application domains of Al-doped ZnO (AZO).

Transparent Conducting Electrodes

TCOs are doped metal oxides used in optoelectronic devices such as flat panel displays and photovoltaic cells. The next-generation eco-friendly optoelectronics is anticipated to be inspired by the design and development of transparent conductive thin films with innovative and superior functions like flexibility and recyclability. Today's electronic and optoelectronic devices such as flat panel displays, light-emitting diodes, photovoltaic cells and smart phones, *etc.* encompass transparent conductors as one of the key elements. Current technology is reliant on the use of Indium Tin Oxide (ITO) as a transparent conducting material. Even though ITO offers numerous exceptional features like large transmission and low resistance, it still faces certain deficiencies like mechanical flexibility and high-temperature processing conditions. The industry standards in TCOs suggest the use of ITO. This material has a low resistivity of the order of 10^{-4} $\Omega \cdot$cm and a transmittance nearly or greater than 80%. The biggest challenges with ITO are that it is scarce and expensive to produce. These issues have been posing a persistent challenge before the scientific community to look for an alternate material that is cost effective with uncompromised properties parallel to ITO.

The applications of TCO coatings have prompted enormous research on their deposition and characterization methods. Various TCO films are currently being applied in optoelectronics including touch panels, electroluminescent, plasma, and field emission displays. In addition, these coatings are also used as heat reflective mirrors, energy efficient windows, gas sensors, as transparent electrodes in photovoltaic cells, and as fire retarding materials. As transparent conductors, these films find applications in vehicle and aircraft windscreen defrosters.

Thin films of metal oxides like cadmium oxide (CdO) and indium oxide (In_2O_3), are a well-known classical example for optically transparent and electrically conducting materials. Thin films of CdO, produced by thermal oxidation of sputtered cadmium were first reported by Badeker *et al.* [50]. Thin films (~100-200 Å) of metals such as Ag, Mg, Cu, Fe, *etc.* were also reported to have similar properties. These films, in general, are not very stable and sensitive to the environment, and their optical and electrical properties change with time. On the other hand, some metal oxide coatings were widely applied because of their stability and hardness which are superior to those of thin metallic films. Other examples of widely used TCOs include Sn doped In_2O_3, F doped SnO_2 (FTO) and Al doped ZnO (AZO). Within this class of materials, there exist specialized textured TCOs that perform not only as transmissive front electrodes and diffusion barriers but also as light trapping units. These units feature widespread light scattering sites at the interface and/or inside of the coatings that prolong light paths with various refractive indices. The simple processing of AZO films brought it to the level of front running material as TCO.

In sol-gel derived AZO films, the concentration, aging time and annealing temperature govern the various features of the films [47]. Recently, D. T. Speaks [48] have shown that the longer aging time reduces the transparency of the films. Also, there is an observation that an optimum concentration of the sol is required to produce the particular crystallite size in the films. Birkholz *et al.* showed that the sharpness of orientation distribution reduces the electrical resistivity of AZO films deposited by reactive DC magnetron sputtering from a metallic Zn-Al (2 wt. %) alloy target [49]. Gardeniers *et al.* demonstrated the relationship between the piezoelectric strain constant and the c-axis orientation of ZnO films prepared by reactive RF magnetron sputtering [51]. For AZO films of certain thickness grown by various types of magnetron sputtering with an AZO target, it was obtained that AZO films deposited by DC magnetron sputtering had high optical mobility, corresponding to high in-grain carrier mobility, high carrier concentration, and a large contribution of grain boundary scattering to carrier transport compared to AZO films deposited by RF magnetron sputtering. With such a high focus on extensive research, AZO systems is an evidence that despite high maturity, progress is underway.

Solar Cells/Photovoltaics

Heterojunction solar cells with an integral conducting transparent layer offer the possibility of fabrication of low-cost solar cells with performance characteristics suitable for large-scale terrestrial applications. The conducting transparent film permits the transmission of solar radiation directly to the active region with little or no attenuation. Li *et al.* have reported the preparation of ZnO thin films where the additives are seen to dramatically influence the electrical and optical properties of as-deposited films, in which, the ultra large haze factor of 78% across the visible spectrum and competitive resistivity with high carrier mobility of 22.92 cm^2 V^{-1} s^{-1} were achieved as compared to undoped ZnO [52]. Furthermore, the combination of Al dopant and additives into the precursor could successfully prove to enhance transmittance (> 83%), lower the resistivity ($3.49X10^{-3}$ Ω cm) and augment the haze value (> 62%) in AZO, which presents an enormous potential for the application of front transparent electrode in Si-based thin film solar cells with superior light scattering capability and potential cost-effectiveness for sustainable production.

Loo *et al.* have demonstrated that the atomic layer deposited ZnO films can provide excellent passivation of lightly-doped n- and p-type c-Si surfaces [53]. It was found that the presence of a high-quality SiO_2 interface layer between c-Si and ZnO was key in achieving high levels of surface passivation. Moreover, it was required to cap ZnO by Al_2O_3 to obtain surface passivation. It was established that the level of surface passivation improved with increasing electron densities in the ZnO films. The addition of extrinsic dopants (Al) was found to reduce the work function of ZnO, which led the enhanced field-effect passivation by reducing the hole concentration at the c-Si surface. The AZO films described in their work were highly transparent, conductive, suitable as anti-reflection coating, and moreover provide high levels of surface passivation. Due to these properties, their study on AZO films provides new opportunities for c-Si solar cells. For instance, the AZO films could potentially serve as an alternative to SiNx as a passivating antireflection coating at the front surface of solar cells. In this way, it would loosen the lateral conduction requirements of the emitter or omit the use of firing-through metallization. Moreover, the passivation of AZO films is of specific interest as a highly transparent passivating electron contact. For this application to be successful, a sufficiently low contact resistivity of passivating AZO with c-Si would be a prerequisite. In the recent work on AZO coatings by RF driven magnetron sputtering, resistivity below 3×10^{-3} Ωcm has been obtained with an averaged transmittance above 90%. In the same work, no intentional substrate heating is noticed [54]. These studies are encouraging the current research community to adopt AZO as chief materials for surface passivation and antireflection coating suitable for solar cells.

Flexible/Organic/Smart Electronics

Flexible and large surface area electronics have become increasingly important areas of research. The underlying reason of this rise in flexible electronics is the demand for enabling a natural interaction between electronics and humans. Large-area, flexibility, conformability and stretchability are being achieved because of the development of some strongly influential polymers and organic compounds. In the displays based on organic light emitting diodes, thin-film transistors (TFTs) with an active layer of semiconductor metal oxide are popularly used. Simultaneously there is an evolution of a ferroelectric field effect transistor (FeFET) which is becoming an integral part of flexible electronics. This 'flexibility' puts a natural pre-condition on the material to be susceptible to the imposition of strain without causing any unfavorable defect formation. Recently, $Ba_{0.5}Sr_{0.5}TiO_3$ (BSTO) and ferroelectric $BaTiO_3$ (BTO) epitaxially grown on a flexible muscovite substrate have been widely used for their superior nonlinear dielectric behaviors. The incorporation of AZO semiconductor layer in a FeFET (Fig. **6**) designed out of BSTO and BTO has recently given an assurance to develop a mechanically controllable transistor which is a huge value addition to flexible electronics [55].

A facile fabrication of AZO nanoflakes vertically grown on flexible polymer substrates with enhanced light scattering and dye loading capabilities has been found to significantly enhance the power conversion efficiency of flexible dye-sensitized solar cells. Particularly, AZO nanoflake photoanodes of 6 μm thick was responsible for this enhancement in efficiency due to the increased electron mobility and reduced carrier recombination [56]. AZO films have a significant impact on technological applications of passive and active devices where semiconductor-metal transition (SMT) is an important key player in governing the features of the devices. The electro-thermal control was demonstrated when AZO was incorporated in a multilayered thin film system based on vanadium dioxide (VO_2). The small electric field was found to be responsible for enabling SMT in VO_2 based systems. AZO film could facilitate the good control of the SMT of the VO_2 thin film and its associated properties [57]. Hence, it is established that the smart coatings offer a great space to the AZO films.

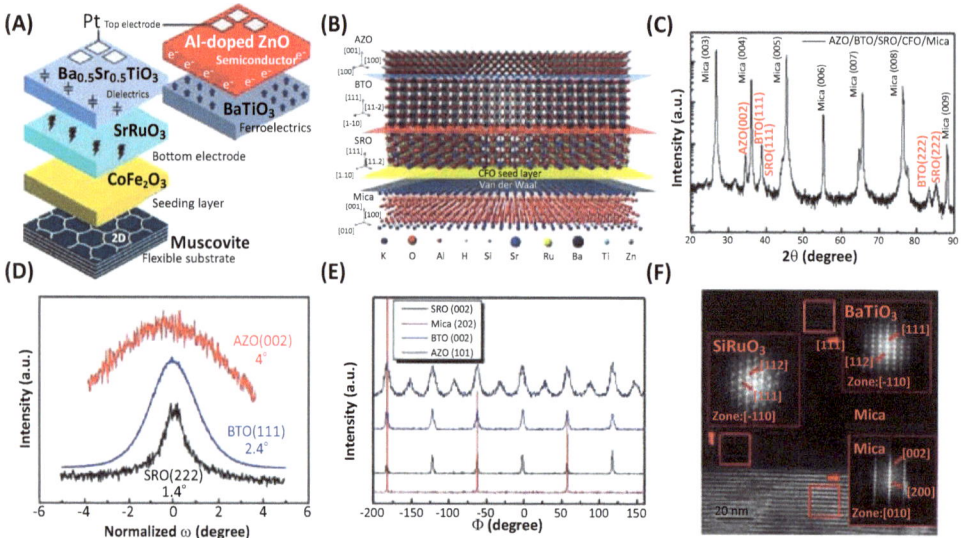

Fig. (6). Structural characteristics of heteroepitaxy, where AZO is used to develop FeFET **(A)** Schematic of BSTO and BTO systems. **(B)** Schematic of the epitaxial relationship. **(C)** Out-of-plane x-ray 2θ-θ scan of the heterostructure. a.u., arbitrary units. **(D)** Rocking curves of SRO(222), BTO(111), and AZO(002). **(E)** Φ-Scan of muscovite{202}, SRO{002}, BTO{002}, and AZO{101}. **(F)** Cross-sectional TEM image at the interface and the corresponding fast Fourier transform (FFT) patterns in the insets. [**Credit:** D. L. Ko *et al.*, Science Advances (2020). DOI:10.1126/sciadv.aaz3180].

Plasmonics

Recently a new field of research namely 'plasmonics' is taking a great momentum in science and technology [58]. The field is connected to surface Plasmon resonance (SPR). In SPR, the interaction of light with free electrons at a metal–dielectric interface takes place. The photons transfer their energy to cause a collective excitation of free electrons. This collective oscillation of free electron gas at the interface is called a surface plasmon. The transfer of energy happens only at a given wavelength of light particularly when the photon momentum coincides with that of plasmon.

Metals like Ag, Au and Pt, *etc.*, which have been popular for SPR, are seen to seriously limit the optical performance of a device. AZO has emerged as a new plasmonic material that has experimentally demonstrated negative refractive index in an AZO/ZnO interface in the near-infrared range. Hence, AZO could overcome the limitation of traditional metals being used to excite SPR. This makes AZO a better optical metamaterial to enable more efficient devices [59]. Simulation studies noticed the localized surface –plasmon-polariton (SPP) resonance in the holographically patterned AZO structures which could establish the use of AZO films for plasmonic devices [60]. Another theoretical study on

AZO confirmed the presence of negative permittivity at IR wavelengths. This was found to be happening because of an increased concentration of free electrons in ZnO system by Al-doping [61]. Such results offer a great scope for the design of plasmonic imaging devices. Further, systematic simulation results on the total light absorption (TLA) in AZO square lattice hole array/dielectrics/metal stacks in the plasmonic regime of AZO showed the evidence of SPP and Fabry-Perot (F-P) resonance which has a great potential to be applied to understand and develop plasmonic lasers. The evidence of SPP and Fabry-Perot (F-P) resonance is demonstrated here in Fig. (7). The thickness of AZO films was found to influence the permittivity in both the visible and infrared regions, and the reason behind was the change in crystallinity with thickness. Such results encouraged the feasibility of the AZO absorber at 2–5 μm [62]. Overall, the AZO based coatings are still full of opportunities to carry out systematic experimental investigations in the direction of achieving more practical applications in the plasmonic domain.

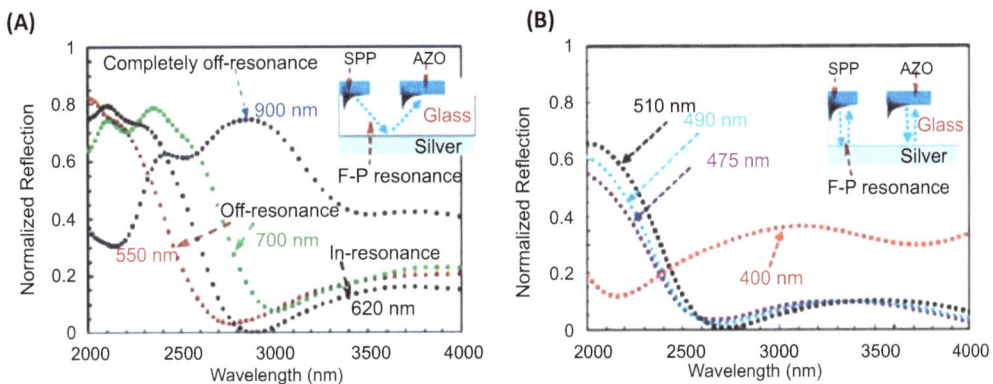

Fig. (7). (a) Simulated reflection spectra from a stack of AZO hole arrays/spacer/silver for a spacer thickness of 550, 620, 700 and 900 nm. Total light absorption (TLA) is achieved when a surface plasmon polariton (SPP) couples with the Fabry-Perot resonance in zig-zag fashion, as shown in the inset; **(b)** Simulated reflection spectra for a spacer thickness of 475, 490, and 510 nm. [**Credit:** D. George *et al.* Photonics 2017, 4, 35; doi:10.3390/photonics4020035].

CONCLUSION

ZnO based films deposited by any of the existing methods are found to show excellent behaviour as TCO. Particularly, AZO films are of great importance from both viewpoints of fundamental materials' behaviour on the structural factors limiting carrier transport and practical technology, and for achieving the carrier mobility required by the applications including electrical and optoelectronic devices. Despite greater developments, there is a huge scope for further augmentation of features suitable to optoelectronic industries. AZO presents enormous potential for the application of front transparent electrode in Si based thin film solar cells with superior light-scattering capability and potential cost-effective for sustainable production. The emerging fields like flexible electronics and plasmonics are looking at AZO as a significant future material. Also, AZO seems to be a strong competitor to its peers like ITO in catering to the commercial entities based on transparent electronics such solar photovoltaics. The comparative analysis presented here gives a clear insight into the applications of AZO in various domains and concludes that on one side, it has maturity in the TCO domain and on the other side, it has a huge opening in plasmonics.

CONSENT FOR PUBLICATION

Not applicable.

CONFLICT OF INTEREST

The authors declare no conflict of interest, financial or otherwise.

ACKNOWLEDGEMENT

Declared none.

REFERENCES

[1]   Craddock, P. T., 2000 years of zinc and brass, British Museum 1998. ISBN 978-0-86159-124-4.

[2]   Moezzi A, McDonagh AM, Cortie MB. Zinc oxide particles: Synthesis, properties and applications. Chem Eng J 2012; 185-186: 1-22.
[http://dx.doi.org/10.1016/j.cej.2012.01.076]

[3]   Brown HE. Zinc Oxide Rediscovered. New York, NY, USA: The New Jersey Zinc Company 1957.

[4]   Leong ESP, Yu SF, Chong MK, Tan OK, Pita K. Metal-oxide-SiO/sub 2/ composite ZnO lasers. IEEE Photonics Technol Lett 2005; 17(9): 1815-7.
[http://dx.doi.org/10.1109/LPT.2005.853010]

[5]   Krishnamoorthy S, Iliadis AA. Properties of high sensitivity ZnO surface acoustic wave sensors on SiO2/(100) Si substrates. Solid-State Electron 2008; 52(11): 1710-6.
[http://dx.doi.org/10.1016/j.sse.2008.06.039]

[6]   Kumawat A, Chattopadhyay S, Misra KP, Halder N, Jain SK, Choudhary BL. Blue-shift in the optical band gap of sol-gel derived $Zn_{(1-x)}Sr_xO$ nanoparticles. Solid State Sci 2020; 108: 106379.
[http://dx.doi.org/10.1016/j.solidstatesciences.2020.106379]

[7]   Subramanian V, Bakhishev T, Redinger D, Volkman SK. Solution-Processed Zinc Oxide Transistors for Low-Cost Electronics Applications. J Disp Technol 2009; 5(12): 525-30.
[http://dx.doi.org/10.1109/JDT.2009.2029124]

[8]   Dietl T, Ohno H, Matsukura F, Cibert J, Ferrand D. Zener model description of ferromagnetism in zinc-blende magnetic semiconductors. Science 2000; 287(5455): 1019-22.
[http://dx.doi.org/10.1126/science.287.5455.1019] [PMID: 10669409]

[9]   Wang HT, Kang BS, Ren F, *et al.* Hydrogen-selective sensing at room temperature with ZnO nanorods. Appl Phys Lett 2005; 86(24): 243503.
[http://dx.doi.org/10.1063/1.1949707]

[10]   Han J, Fan F, Xu C, *et al.* ZnO nanotube-based dye-sensitized solar cell and its application in self-powered devices. Nanotechnology 2010; 21(40): 405203.
[http://dx.doi.org/10.1088/0957-4484/21/40/405203] [PMID: 20829568]

[11]   Cao Y, Miao L, Tanemura S, Tanemura M, Kuno Y, Hayashi Y. Low resistivity p-ZnO films fabricated by sol-gel spin coating. Appl Phys Lett 2006; 88(25): 251116.
[http://dx.doi.org/10.1063/1.2215618]

[12]   Misra KP, Shukla RK, Srivastava A, Srivastava A. Blueshift in optical band gap in nanocrystalline $Zn_{1-x}Ca_xO$ films deposited by sol-gel method. Appl Phys Lett 2009; 95(3): 031901.
[http://dx.doi.org/10.1063/1.3184789]

[13]   Srivastava A, Kumar N, Misra KP, Khare S. Blue-light luminescence enhancement and increased band gap from calcium-doped zinc oxide nanoparticle films. Mater Sci Semicond Process 2014; 26: 259-66.
[http://dx.doi.org/10.1016/j.mssp.2014.05.001]

[14]   Tripathi A, Misra KP, Shukla RK. UV enhancement in polycrystalline Ag-doped ZnO films deposited by the sol–gel method. J Lumin 2014; 149: 361-8.
[http://dx.doi.org/10.1016/j.jlumin.2013.12.043]

[15]   Misra KP, Srivastava A, Shukla RK, Misra P, Srivastava A. Polarization Characteristics Variation of Visible Light on Reflection from ZnO Based Amorphous Films. Jpn J Appl Phys 2010; 49(6): 062602.
[http://dx.doi.org/10.1143/JJAP.49.062602]

[16]   Kumar N, Misra KP, Jain SK, Choudhary BL. Structural and morphological properties of Ce doped ZnO. AIP Conf Proc 2013; 1536(1): 605-6.
[http://dx.doi.org/10.1063/1.4810372]

[17]   Chattopadhyay S, Misra KP, Agarwala A, Rao A, Babu PD. Correlated quartic variation of band gap and NBE energy in sol-gel derived Zn1−Co O nanoparticles. Mater Chem Phys 2019; 227: 236-41.
[http://dx.doi.org/10.1016/j.matchemphys.2019.02.003]

[18]   Srivastava A, Shukla RK, Misra KP. Phtotoluminescence from screen printed ZnO based nanocrystalline films. Cryst Res Technol 2011; 46(9): 949-55.

[19]   Shukla R K, Srivastava A, Pandey A C, Misra K P, Pandey M. Sensitivity of polyaniline-zinc oxide composite to humidity. 2015; 188(5): 26.

[20]   Sharma A, Khangarot RK, Kumar N, Chattopadhyay S, Misra KP. Rise in UV and blue emission and reduction of surface roughness due to the presence of Ag and Al in monocrystalline ZnO films grown by sol-gel spin coating. Mater Technol 2021; 36(9).
[http://dx.doi.org/10.1080/10667857.2020.1776029]

[21]   Chattopadhyay S, Misra KP, Agarwala A, *et al.* Dislocations and particle size governed band gap and ferromagnetic ordering in Ni doped ZnO nanoparticles synthesized *via* co-precipitation. Ceram Int 2019; 45(17): 23341-54.
[http://dx.doi.org/10.1016/j.ceramint.2019.08.034]

[22]   Misra KP, Jain S, Agarwala A, Halder N, Chattopadhyay S. Effective Mass Model Supported Band Gap Variation in Cobalt-Doped ZnO Nanoparticles Obtained by Co-Precipitation. Semiconductors 2020; 54(3): 311-6.
[http://dx.doi.org/10.1134/S1063782620030136]

[23]   Ramamoorthy K, Arivanandhan M, Sankaranarayanan K, Sanjeeviraja C. Highly textured ZnO thin films: a novel economical preparation and approachment for optical devices, UV lasers and green LEDs. Mater Chem Phys 2004; 85(2-3): 257-62.
[http://dx.doi.org/10.1016/j.matchemphys.2003.09.018]

[24]   Dhananjay , Krupanidhi SB. Low threshold voltage ZnO thin film transistor with a $Zn_{0.7}Mg_{0.3}O$ gate dielectric for transparent electronics. J Appl Phys 2007; 101(12): 123717.
[http://dx.doi.org/10.1063/1.2748863]

[25]   Coleman VA, Jagadish C. Basic Properties and Applications of ZnO. Elsevier 2006.
       [http://dx.doi.org/10.1016/B978-008044722-3/50001-4]

[26]   Meyer J, Görrn P, Hamwi S, Johannes HH, Riedl T, Kowalsky W. Indium-free transparent organic
       light emitting diodes with Al doped ZnO electrodes grown by atomic layer and pulsed laser deposition.
       Appl Phys Lett 2008; 93(7): 073308.
       [http://dx.doi.org/10.1063/1.2975176]

[27]   Minami T, Sato H, Nanto H, Takata S. Group III impurity doped zinc oxide thin films prepared by RF
       magnetron sputtering. Jpn J Appl Phys 1985; 24(Part 2, No. 10): L781-4.
       [http://dx.doi.org/10.1143/JJAP.24.L781]

[28]   Minami T. New n-type transparent conducting oxides. MRS Bull 2000; 25(8): 38-44.
       [http://dx.doi.org/10.1557/mrs2000.149]

[29]   Shukla RK, Srivastava A, Srivastava A, Dubey KC. Growth of transparent conducting nanocrystalline
       Al doped ZnO thin films by pulsed laser deposition. J Cryst Growth 2006; 294(2): 427-31.
       [http://dx.doi.org/10.1016/j.jcrysgro.2006.06.035]

[30]   Tominaga K, Manabe H, Umezu N, Mori I, Ushiro T, Nakabayashi I. Film properties of ZnO:Al
       prepared by cosputtering of ZnO:Al and either Zn or Al targets. J Vac Sci Technol A 1997; 15(3):
       1074-9.
       [http://dx.doi.org/10.1116/1.580432]

[31]   Kim KH, Park KC, Ma DY. Structural, electrical and optical properties of aluminum doped zinc oxide
       films prepared by radio frequency magnetron sputtering. J Appl Phys 1997; 81(12): 7764-72.
       [http://dx.doi.org/10.1063/1.365556]

[32]   Addonizio ML, Antonaia A, Cantele G, Privato C. Transport mechanisms of RF sputtered Al-doped
       ZnO films by H$_2$ process gas dilution. Thin Solid Films 1999; 349(1-2): 93-9.
       [http://dx.doi.org/10.1016/S0040-6090(99)00186-8]

[33]   Kon M, Song PK, Shigesato Y, Frach P, Mizukami A, Suzuki K. Al-doped ZnO films deposited by
       reactive magnetron sputtering in mid-Frequency mode with dual cathodes. Jpn J Appl Phys 2002;
       41(Part 1, No. 2A): 814-9.
       [http://dx.doi.org/10.1143/JJAP.41.814]

[34]   Hu J, Gordon RG. Textured aluminum-doped zinc oxide thin films from atmospheric pressure
       chemical-vapor deposition. J Appl Phys 1992; 71(2): 880-90.
       [http://dx.doi.org/10.1063/1.351309]

[35]   Hiramatsu M, Imaeda K, Horio N, Nawata M. Transparent conducting ZnO thin films prepared by
       XeCl excimer laser ablation. J Vac Sci Technol A 1998; 16(2): 669-73.
       [http://dx.doi.org/10.1116/1.581085]

[36]   Kim H, Gilmore CM, Horwitz JS, *et al.* Transparent conducting aluminum-doped zinc oxide thin films
       for organic light-emitting devices. Appl Phys Lett 2000; 76(3): 259-61.
       [http://dx.doi.org/10.1063/1.125740]

[37]   Agura H, Suzuki A, Matsushita T, Aoki T, Okuda M. Low resistivity transparent conducting Al-doped
       ZnO films prepared by pulsed laser deposition. Thin Solid Films 2003; 445(2): 263-7.
       [http://dx.doi.org/10.1016/S0040-6090(03)01158-1]

[38]   Tang W, Cameron DC. Aluminum-doped zinc oxide transparent conductors deposited by the sol-gel
       process. Thin Solid Films 1994; 238(1): 83-7.
       [http://dx.doi.org/10.1016/0040-6090(94)90653-X]

[39]   Musat V, Teixeira B, Fortunato E, Monteiro RCC, Vilarinho P. Al-doped ZnO thin films by sol–gel
       method. Surf Coat Tech 2004; 180-181: 659-62.
       [http://dx.doi.org/10.1016/j.surfcoat.2003.10.112]

[40]   Bel Hadj Tahar R. Structural and electrical properties of aluminum-doped zinc oxide films prepared by

sol–gel process. J Eur Ceram Soc 2005; 25(14): 3301-6.
[http://dx.doi.org/10.1016/j.jeurceramsoc.2004.08.028]

[41]　Nomoto J, Inaba K, Osada M, Kobayashi S, Makino H, Yamamoto T. Highly (0001)-oriented Al-doped ZnO polycrystalline films on amorphous glass substrates. J Appl Phys 2016; 120(12): 125302-, 125302-125311.
[http://dx.doi.org/10.1063/1.4962943]

[42]　Nomoto J, Makino H, Yamamoto T. Carrier mobility of highly transparent conductive Al-doped ZnO polycrystalline films deposited by radio-frequency, direct-current, and radio-frequency-superimposed direct-current magnetron sputtering: Grain boundary effect and scattering in the grain bulk. J Appl Phys 2015; 117(4): 045304-, 045304-045309.
[http://dx.doi.org/10.1063/1.4906353]

[43]　Nomoto J, Makino H, Yamamoto T. High-Hall-mobility Al-doped ZnO films having textured polycrystalline structure with a well-defined (0001) orientation. Nanoscale Res Lett 2016. 11, 320.

[44]　Nomoto J, Makino H, Yamamoto T. Correlation between carrier transport and orientation evolution of polycrystalline transparent conductive Al-doped ZnO films. Thin Solid Films 2016; 620: 2-9.
[http://dx.doi.org/10.1016/j.tsf.2016.07.078]

[45]　Nomoto J, Inaba K, Kobayashi S, Makino H, Yamamoto T. Interface layer to tailor the texture and surface morphology of Al-doped ZnO polycrystalline films on glass substrates. J Cryst Growth 2017; 468: 645-9.
[http://dx.doi.org/10.1016/j.jcrysgro.2016.12.072]

[46]　Fujimura N, Nishihara T, Goto S, Xu J, Ito T. Control of preferred orientation for ZnO films: control of self-texture. J Cryst Growth 1993; 130(1-2): 269-79.
[http://dx.doi.org/10.1016/0022-0248(93)90861-P]

[47]　Khan TM, Zakria M, Shakoor RI, Raffi M, Ahmad M. Mechanisms of composite-hydroxide-mediated approach for the synthesis of functional ZnO nanostructures and morphological dependent optical emissions. Adv Mater Lett 2015; 6: 592-9.
[http://dx.doi.org/10.5185/amlett.2015.5876]

[48]　Speaks DT. Effect of concentration, aging, and annealing on sol gel ZnO and Al-doped ZnO thin films. International Journal of Mechanical and Materials Engineering 2020; 15(1): 2.
[http://dx.doi.org/10.1186/s40712-019-0113-6]

[49]　Birkholz M, Selle B, Fenske F, Fuhs W. Structure-function relationship between preferred orientation of crystallites and electrical resistivity in thin polycrystalline ZnO:Al films. Phys Rev B Condens Matter 2003; 68(20): 205414-, 205414-205418.
[http://dx.doi.org/10.1103/PhysRevB.68.205414]

[50]　Baedeker K. Über die elektrische Leitfähigkeit und die thermoelektrische Kraft einiger Schwermetallverbindungen. Annalen der Physik (Leipzig) 1907; 327(4): 749-66.
[http://dx.doi.org/10.1002/andp.19073270409]

[51]　Gardeniers JGE, Rittersma ZM, Burger GJ. Preferred orientation and piezoelectricity in sputtered ZnO films. J Appl Phys 1998; 83(12): 7844-54.
[http://dx.doi.org/10.1063/1.367959]

[52]　Li J, Sathasivam S, Taylor A, Carmalt CJ, Parkin IP. Single step route to highly transparent, conductive and hazy aluminium doped zinc oxide films. RSC Advances 2018; 8(74): 42300-7.
[http://dx.doi.org/10.1039/C8RA09338E] [PMID: 35558400]

[53]　van de Loo BWH, Macco B, Melskens J, Beyer W, Kessels WMM. Silicon surface passivation by transparent conductive zinc oxide. J Appl Phys 2019; 125(10): 105305.
[http://dx.doi.org/10.1063/1.5054166]

[54]　Abdulmunem OM, Mohammed Ali MJ, Hassan ES. Optical and structural characterization of aluminium doped zinc oxide thin films prepared by thermal evaporation system. Opt Mater 2020; 109:

110374.
[http://dx.doi.org/10.1016/j.optmat.2020.110374]

[55]   Mechanically controllable nonlinear dielectrics (2020, March 16) retrieved 2 October 2020 from: https://phys.org/news/2020-03-mechanically-nonlinear-dielectrics.html

[56]   Zhao C, Zhang J, Hu Y, Robertson N, Hu PA, Child D. Desmond Gibson and Yong Qing Fu. Sci Rep 2015; 5: 17750.
[http://dx.doi.org/10.1038/srep17750] [PMID: 26631685]

[57]   Skuza JR, Scott DW, Mundle RM, Pradhan AK. Electro-thermal control of aluminum-doped zinc oxide/vanadium dioxide multilayered thin films for smart-device applications. Sci Rep 2016; 6(1): 21040.
[http://dx.doi.org/10.1038/srep21040] [PMID: 26884225]

[58]   Zalevsky Z, Abdulhalim I. Integrated. William Andrew, USA: Nanophotonic Devices 2014.

[59]   Naik GV, Liu J, Kildishev AV, Shalaev VM, Boltasseva A. Demonstration of Al:ZnO as a plasmonic component for near-infrared metamaterials. Proc Natl Acad Sci USA 2012; 109(23): 8834-8.
[http://dx.doi.org/10.1073/pnas.1121517109] [PMID: 22611188]

[60]   George D, Li L, Jiang Y, *et al.* Localized surface plasmon polariton resonance in holographically structured Al-doped ZnO. J Appl Phys 2016; 120(4): 043109.
[http://dx.doi.org/10.1063/1.4960018]

[61]   Masouleh FF, Sinno I, Buckley RG, Gouws G, Moore CP. Characterization of conductive Al-doped ZnO thin films for plasmonic applications. Appl Phys, A Mater Sci Process 2018; 124(2): 174.
[http://dx.doi.org/10.1007/s00339-018-1600-y]

[62]   Zheng H, Zhang RJ, Li DH, *et al.* Optical Properties of Al-Doped ZnO Films in the Infrared Region and Their Absorption Applications. Nanoscale Res Lett 2018; 13(1): 149.
[http://dx.doi.org/10.1186/s11671-018-2563-9] [PMID: 29752609]

CHAPTER 4

Commercial Applications of Synthetic Fibres

Sunanda Das[1],*

[1] *Department of Chemistry, Chaudhary Mahadeo Prasad Degree College, University of Allahabad, Prayagraj, Uttar Pradesh, India*

Abstract: Man-made fibres are produced from chemical substances known as synthetic fibres. Synthetic fibre or a synthetic polymer made from molecules of monomer joined together to form long chains, is also known as an artificial fibre. Besides polymer-based synthetic fibres, other types of fibres that have special commercial applications and importance. These include the fibers made of carbon, glass, metal and ceramics. Polymer-based synthetic fibres are produced by various processes such as melt spinning, dry spinning and wet spinning.

The melt spinning technique is used to produce polymers such as polyethene, polyetheneterephthalate, cellulose triacetate, polyvinyl chloride, nylon, *etc.* Cellulose acetate, cellulose triacetate, acrylic, modacrylic, polyvinyl chloride and aromatic nylon are artificial fibres manufactured by dry-spinning. In contrast, the wet spinning process is used for aromatic nylon, polyvinyl chloride fibres, acrylic, modacrylic and viscose rayon from regenerated cellulose.

The importance and usefulness of synthetic fibres are because they have enhanced properties compared to natural fibres, which come from plants or animals. Still, each type is valued for different reasons.

Keywords: Artificial fibres, Commercial applications, Polymers, Polyesters, Polyolefins, Polyamides, Polyurethane, Rayons, Synthetic fibres.

INTRODUCTION

Synthetic fibres, like natural fibres, are made up of small units known as monomers which join together and make extensive units called polymers. In comparison, natural fibres like cellulose are obtained from plants and animals. Synthetic fibres are obtained by processing petrochemicals chemically.

Synthetic fibres can be woven into fabrics like natural fibres and have wide-ranging commercial applications (Chart **1**).

---

* **Corresponding author Sunanda Das:** Department of Chemistry, Chaudhary Mahadeo Prasad Degree College, University of Allahabad, Prayagraj, Uttar Pradesh, India; E-mail: sunanda.das@gmail.com

Arti Srivastava, Mridula Tripathi, Kalpana Awasthi and Subhash Banerjee (Eds.)

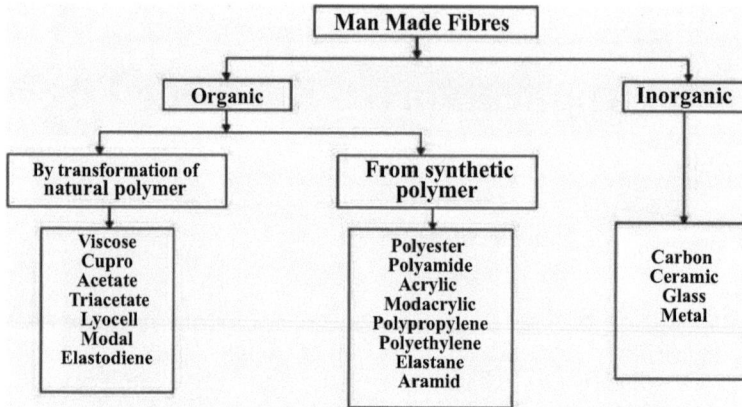

Chart. (1). Classification of synthetically made fibres.

In household articles, synthetic fibres find many uses ranging from ropes, furniture, buckets, and containers to many others. In addition, highly specialized uses are in ships, spacecraft, aircraft, healthcare, *etc.* The manufacturing of synthetic fibres depends upon the types of chemicals used. Fibres differ in their characteristics in their strength, water absorbing capacity, resistance to burning and heat, tensile strength, cost, durability, *etc.* In current scenarios, many synthetic fibres like nylon, rayon, polyester and acrylic are extensively used commercially.

The types, properties and uses of some synthetic fibers are explained in Table **1**.

Table 1. Synthetic fibres: Types, Properties and Uses.

Fiber Type	Flame Properties	Residue/Smell	Characteristic	Use
Synthetic	Melts and retreats from flame, flame burns with bright glow Catches fire easily Easy to melt Burns with dark and thick smoke	Hard, brittle or lumpy sticky beads Harsh acidic smell or blended smell	Raw materials-petroleum origin Do not bleed color and shrink when washed, easy to maintain, wash and dry, less expensive, easily available, more elastic in nature, resistant to insect attack, tough and durable	Warm and protective clothing for extreme climates, fabric for everyday wear, sportswear, fillers, hair wigs, ropes, parachutes, nets, tents, carpets, furnishing, upholstery, utensils,

(Table 1) cont.....

Fiber Type	Flame Properties	Residue/Smell	Characteristic	Use
Rayon	Burns with a steady flame with dark grey to black smoke	gritty and brittle, almost no ash paper or wood-like smell	Made up of chemically treating wood pulp, Known as artificial silk Absorbs moisture and comfortable to wear	Apparels like suits, slacks, jackets,etc Because of its strength used in making automobile tyre cords Mixed with wool to make carpets and blankets Mixed with cotton to make bedspreads, bed sheets, home furnishings
Acetate	Burns and melts sizzle, fiber continues to burn after the flame source is removed.	Hard and black bead. The acidic or vinegary smell.	Soft, smooth, natural feel, high flexibility, high gloss. Effected by acids and bleaches, resistant to alkalies. Effected by sunlight. Low moisture absorbance, high elongation, more sensitive to heat, thermoplastic	Dresses, foundation garments, lingerie, lining material, household furnishings, sportswear, garments where pleat retention is important, certain specialty fabrics, cigarette filter, diapers, surgical products, cellulose acetate ribbons,playing cards, photographic films
Polyester	Melts and burns at the same time, retreats from flame, Black smoke with melted or brown bead	Sweet or acidic smell. Drips while burning, drippings are sometimes with fire	Manufactured from petroleum. A polymer of ester. Extremely strong fiber, resistant to chemicals, less wrinkles, does not absorb water easily, easy to dry less fading, durable	Textiles-dress materials, sarees, upholstery, hosiery Blended with natural fibers(like wool and cotton) Sail of sailboats Water hoses(firefighting purposes) As Mylar for making film and magnetic recording tapes
Polyolefin	Shrinks quickly in flame, burns and melts, burns when flame is withdrawn.	Smell not defined Hard round bead maybe light brown.	Hard and opaque plastic, high thermal stability, heat-ageing resistance, stronger mechanical properties, good resistance to heat, flame, ozone, chemicals, solvents, low water absorption	Insulation of electrical wiring in high temp., lighting fixtures, heating cable for pipe, electrical insulators(radar, television), fuel hose products, air filters, pool filters, battery cover, tank linings, footwear, building products, shrink wrap films, containers, crates, lab equipment

(Table 1) cont.....

Fiber Type	Flame Properties	Residue/Smell	Characteristic	Use
Polyamide	Melts as it burns, flame orange/yellow in color, Inherent resistance to flame combustion,	-	Very strong, hard-wearing, does not rot, low absorbency, dries quickly, good resistance to fatigue Unaffected by commonly used solvents and oils, hydrocarbons, esters, alcohols, Freons Durable, long lasting, inexpensive Resists bacteria, insects Damaged by sunlight –discolours it, makes it weak, affected by alkaline or highly acidic environment Not comfortable to wear	Clothing, ropes seatbelts, carpets and rugs, furnishing material
Acrylic	Melts and burns rapidly with spluttering flame, retreats from flame	Turns into beads, hard brittle irregular shaped ash Pungent acidic smell	Lightweight, soft, warm like wool Can be dyed in variety of colors Shrink resistant, wrinkle resistant, cheaper than wool Storage is easier than wool (moth resistant)	Wollen clothes-sweaters, hats, scarves, gloves, garment materials Home furnishing fabric, As fake fur for toys and fur accesories
Nylon	Melts then burn, burns slowly	Burns with white smoke, turns into brown bead On burning smells like vegetables or plastic	Made up of coal, petroleum oil, air and water First synthetic fiber made entirely from chemicals Strong, elastic, wrinkle free, absorbs little water, easy to wash	Ropes for rock climbing, fishing nets, tents, parachutes, home furnishings, sleeping bags Clothes (including socks) Toothbrushes, seat belts
Aramid	Inherent flame resistance, does not melt	-	Low density and high strength, light weight resistance to fatigue	Ropes, cables, aerospace, auto applications

(Table 1) cont.....

Fiber Type	Flame Properties	Residue/Smell	Characteristic	Use
Kevlar	Flame retardancy	-	-	Fire proofing walls, insulation of electrical wires, cut-resistant gloves, helmets, bullet-proof wests, astronauts suits, Sports items (rackets), reinforced tyres, bicycles, hoses, belts, pars of aircraft, ship hulls, driving gloves, walking boots
Polyvinyl	High melting point	-	Non-reactive to chemicals, nontoxic, excellent corrosion resistance, biologically degradable, soluble in water, insoluble in organic solvents	Medical applications (blood storage containers). In transport vehicles(dash board, door panels Building and construction material, Packaging, pipes and wire coatings, toys, consumer goods, clothings, non-food packaging, plumbing and electric cable insulation
Nomex	Excellent heat and flame resistant, self-extinguishing,does not melt or drip, does not break open after fire exposure	-	Good textile properties, good dimensional stability, high levels of electrical, chemicaland mechanical integrity, low weight, high tenacity, anti-static, can be washed at 95 degrees or dry cleaned, low abrasion and high tear strength	Gloves, nomex hood, fire fighter protection, drivers protection, driving shoes, protective shoes, filtering material, nomex composites
Spandex	Melts, does not retreat from flame, Black or dark ash	an acidic or rubber smell like hot pencil eraser	Very stretchy, keeps its shape, lightweight, resistance to damage by sun, not damaged by perspiration, lightweight, garments comfortable to wear	Sportswear, swim suits, rowing suits, cycling jersey, undergarments, socks, slacks, jogging and exercising suits, T shirts, stretchable jeans Medical uses- surgical hose, orthopedic braces Home furnishings

Following are details of some of the most commonly used synthetic fibres:

RAYON

These are the most widely used semi-synthetic fibres. Rayon or artificial silk is a manufactured fibre obtained from a natural source *i.e.,* wood pulp. Rayon, acetate and lyocell are all regenerated cellulose fibres. They originate from the chemical treatment of raw materials, mainly cotton fibres too short to spin into yarns. Rayon can be mixed with cotton or wool, dyed into a wide variety of colours (Figs. **1** & **2**).

Fig. (1). Rayon threads.

Fig. (2). Rayon fabric.

The generic classification of 'Rayon' includes several variants-such as viscose, cuprammonium (cupro) and acetate rayons. Viscose rayon is the commonest of all. The manufacturing process for rayon, acetate and lyocell fibres has technical differences. The basic procedure for manufacturing of rayon and lyocell is with cotton linter or wood chips, which are treated with chemicals and are reduced to a cellulosic solution, which then passes through spinnerets and dries to become mostly pure cellulose filaments. The best thing during the manufacture of lyocell is that chemicals can be recovered and recycled, which makes it the most environmentally friendly fibre to produce. In contrast, rayon and acetate involve many harmful by-products.

Rayon is described as a regenerated fibre because the cellulose obtained from soft woods or linters, which are soft fibres that generally adhere to cotton seeds, is converted into liquid.

In the production of acetate fibre, wood pulp or cotton linters are used, and after chemical treatments, the fibres produced are no longer pure cellulose but form cellulose acetate. Cellulose acetate fibre was the first thermoplastic fibre [1] that melted and softened on heating.

Viscose Rayon

Cellulose from cotton plants is dissolved in a mixture of sodium hydroxide and solvent carbon disulphide, and a viscous liquid called viscose is obtained. This thick liquid is then forced through spinnerets with small holes into an acid bath; thus filaments of viscose rayon are obtained (Fig. **3**).

Fig. (3). Viscose fabric.

Cuprammonium Rayon (Cupro)

Treating copper sulphate solution with an excess of ammonia solution gives a deep blue solution of compound cuprammonium hydroxide. In this solution, cellulose is dissolved, and the resulting solution is forced through spinnerets into an acid bath to produce filaments of cuprammonium rayon (Fig. **4**).

Fig. (4). Cuprammonium rayon.

Acetate Rayon

Cellulose is dissolved in a mixture of acetic acid, sulphuric acid and other chemicals. The solution is then forced through spinnerets to obtain filaments of acetate rayon (Fig. **5**).

Fig. (5). Acetate rayon, sewinglsCool.com.

Sir Joseph Wilson Swan, a British Chemist in 1884 and in1885, first took a practical step by treating cotton cellulose with nitric acid. He prepared highly inflammable fibre made up of nitrocellulose. But later, Swan did not follow up the demonstration of his invention.

Chardonnet silk, was an early type of rayon and the first commercially produced manufactured fibre by Hilaire de Chardonnet (Fig. **6**). Its commercial production began in 1891 at Besancon.

Fig. (6). Hilaire de Chardonnet, Source: sciencephoto.com.

Another type of cellulosic fibre which is most commonly made today, was earlier produced in 1891 by three British chemists by dissolving cellulose xanthate in a dilute solution of sodium hydroxide. This created a yellow, syrupy sulfurous-smelling liquid.

Later in 1905, Samuel Courtauld & Company, a British silk firm, produced viscose or viscose-rayon. Finally, American Viscose Corporation began US commercial production of viscose in 1911.

LYOCELL

Lyocell fiber [2] has a highly crystalline structure. The advantages of lyocell over cotton and viscose rayon are many, as it has twice dry strength and nearly three times the wet strength, which make lyocell water washable. Furthermore, it shows stronger bonding with latexes and with synthetics. In addition, it produces lighter fabrics; shrinkage is less and has higher absorbency, which makes it better than cotton and viscose rayon.

Due to its excellent properties, it is widely used in clothing, non-woven conveyor belts, industrial filter material, and in the medical field. In addition, this fibre can be blended [3] with cotton, hemp, silk or even synthetic fibres, and viscose fibre for textile manufacturing.

Some explorations by scientists all over the world are going on with lyocell [4]. However, experiments for the production of carbon fibre, especially paper and tire cords using lyocell fibre [5, 6] as a raw material are still in process.

Environmental Impact

Rayon production has declined through the years because of environmental concerns caused by the release of carbon disulfide in the air and other by-products into river streams. Ecological impacts and strict environmental pollution regulations in the US have led many American manufacturers to discontinue rayon production. This has led to the development of lyocell by dissolving wood cellulose in nontoxic amine oxide as a solvent. The solvent was washed off, recovered, and reused. Hence, this turned out to be the best solvent, leading to the commercial success of cellulose fibre under the trade name **TENCEL** by Courtaulds in 1994.

POLYESTER

Polyester is more commonly referred to as polyethene terephthalate (PET), a category of polymers which contains ester as a functional group in the main chain. Polyester is a synthetic polymer. It is made up of purified terephthalic acid (PTA) or its dimethyl ester, *i.e.,* dimethyl terephthalate (DMT) and monoethylene glycol (MEG) (Fig. 7).

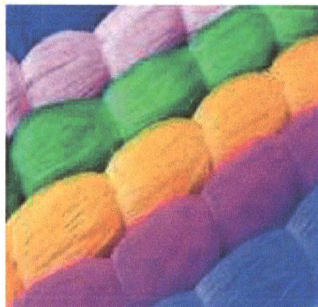

Fig. (7). Polyester fibres.

Polyester fibres are formed from a chemical reaction between an acid and alcohol. It is a long-chain polymer chemically composed of at least eighty five per cent by weight of an ester,a dihydric alcohol, and terephthalic acid. It means it is the linkage of several esters within the fibre. The reaction of alcohol with carboxylic acid results in the formation of esters.

Making Polyester Fibre: The process of creating polyester fibre begins with reacting ethylene glycol with dimethyl terephthalate at high heat. This reaction results in a monomer. The monomer is then reacted with dimethyl terephthalate again to create a polymer. Next, the molten polyester polymer is extruded from the reaction chamber in long strips, and these strips are allowed to cool and dry, then broken into small pieces. The resulting chips are then melted again to create a honey-like substance extruded through a spinneret to produce fibres. These resulting polyester filaments may be cut or reacted with various chemicals to achieve the appropriate end result.

Polyester may include naturally available materials such as cutin of plant cuticles. Natural polyesters and a few synthetic ones are biodegradable, but most synthetic polymers are not. Depending on the chemical structure, polyester can be thermoplastic or thermosetting. The most common polyesters are thermoplastics, which may change shape with heat application.. These fibres have high tenacity, low water absorption, and minimal shrinkage compared to other commercially produced fibres. Thermosetting polyesters are generally unsaturated polymers that are irreversibly hardened by heat. This polymer is a liquid or soft solid that becomes rigid due to strong covalent cross-links.

W.H. Carothers headed the early research on polymer fibresby making nylon, one of the first man-made synthetic fibres (Figs. **8** & **9**). Carothers mixes ethylene glycol and terephthalic acid, but his research was incomplete. It was revived by two British scientists, Whinfeild and Dickson, who patented PET or PETE in 1941. It forms the basis for synthetic fibres like Dacron, terylene and polyester.

Fig. (8). Wallace Hume Carothers.

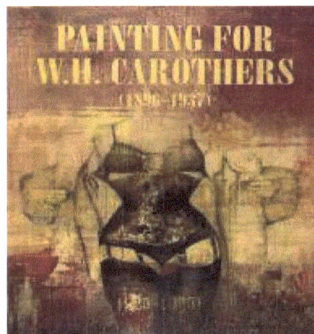

Fig. (9). Source: MutualArt.com.

Polyester is often used in blended fibres. Blended fibres are formed by mixing natural and synthetic fibres. Polywool is made by mixing polyester and wool. Polycot is made by mixing polyester and cotton. Terrycot is made by mixing terylene and cotton. Terry wool is made up of terylene and wool, and Terrysilk is made by mixing terylene and silk.

Ethylene Polyester, also known as **PET(polyethylene terepthalate)** is the most commonly produced and popular form of polyester fibre in the market. The word 'polyester' is synonymous with PET, even though other types of polyester exist. The primary component of PET is petroleum-derived ethylene. In creating polyester fibres, ethylene is the polymer that interacts with other chemicals to develop stable fibrous compounds (Figs. **10** & **11**).

Fig. (10). Polyethylene terepthalate (PET) fibres.

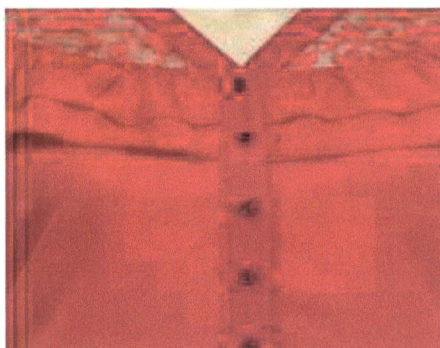

Fig. (11). Polyester dress.

PCDT Polyester – The process of creating PCDT **(poly-1, 4-cyclohexylen--dimethylene terephthalate)** polyester is similar to the process of creating PET polyester. Still,this polyester variant has a different chemical structure, while PCDT consists of ethylene glycol reacted with dimethyl terephthalate. In addition, other production processes make these two-polyesters different. As a result,

PCDT polyester is not as popular as PET, but it is more elastic, which makes it ideal for specific applications. PCDT polyester is also more durable than PET polyester, so these fibres are frequently preferred for heavy-duty applications like upholstery and curtains.

Plant-based Polyesters are made from ethylene glycol reacted with dimethyl terephthalate. While the source of ethylene used in PET and PCDT polyester is petroleum, for plant-based polyester, ethylene sources like cane sugar are used instead. The main advantage of this fibre is that it is biodegradable; however, it costs higher and may be less durable than PET or PCDT.

Commercial Uses of Polyester

It is widely used in the clothing industry, like polyester shirts, jackets, trousers, hats, and polyester yarns. In addition, it is employed in manufacturing of many home furnishing materials like bedsheets, blankets, curtains, pillowcases, *etc.* It's also used in upholstery (Fig. **12**).

Fig. (12). Recycled polyester, Source: dreamstime.com.

An important application of this polymer is in the manufacturing computer mouse pads. In addition, this polymer plays a vital role in the manufacturing of certain types of car tyre reinforcements. Conveyor belts made up of polyester fabric are not uncommon. In automobiles, to protect passengers from accidents, safety belts are often made up of polyester fibres along with other components. In addition, polyester insulating properties, softness and lightweight, this material is used to produce cushioning materials which are used in upholstery padding.

As polyester fibres are abrasion resistant, they can be ideal to use as stain-resistant clothing. In addition, a liquid-crystal display (LCD) is often manufactured with the help of certain polyesters.

Polyesters are employed in the manufacture of bottles and tarpaulin. These fibres can be used to manufacture dielectric films, which are used for capacitors. In addition, polyesters can be used as film insulation in wires and insulating tapes.

Due to thinning properties of polyesters, also known as thixotropic properties, they can be employed as a spray-based polymer that can rapidly fill up wood grains, giving high-quality finishes for certain open-grain timbers. In addition, polyesters can be used for polishing to obtain a glossy and durable finish for musical instruments and other woody articles.

Nowadays, there are two primary types of polyesters- PET (polyethylene terephthalate) and PCDT (poly-1,4-cyclohexylene-dimethylene terephthalate). PET, is the more popular type and has a wide variety of uses. It is stronger than PCDT, though PCDT is more elastic and resilient.

Environmental Impact

Many environmental concerns are there about polyester fibres which cause microplastic pollution in freshwater and saltwater habitats. Polyester, in general, harms the environment. This fibre has unfortunate environmental impacts from production to use and disposal at every stage. Transforming ethylene into polyethylene terephthalate fibres produces more harmful synthetic byproducts. Treatment processes and dyes used by manufacturers make their way into the environment and poison the area's ecosystem.

Commercial Application of Polyester

Polyester fibres [7] take a leading position among all synthetic fibres due to the presence of aliphatic and aromatic parts in the molecular chains and the structure. PET is the predominant polyester used for fibre production not only because of its excellent properties but mainly because of the ease of physical and chemical modifications. In addition, polyester fibre wool has good elasticity and toughness and can be used as a filler for sound absorption.

Polyester fibres can be produced as staple fibres suitable for blending with cotton. These fibres are characterized by low moisture absorption ability to accumulate electrostatic charge on the surface. In addition, due to the high crystalline and dense structure and the presence of benzene rings, these fibres can be dyed at a high temperature up to 140 degrees and elevated pressures.

Polyester fibres are used in fibre-reinforced concrete for industrial, warehouse floors and pavements [8]. These macro and microfibres are used in concrete to provide superior resistance [9] and enhanced toughness. The primary applications of polyester fibres are for concrete or unreinforced concrete, flatwork, footings, foundations, walls, tanks, concrete pipes, burial vaults and pre-stressed beams. In addition, polyester fibre battings, also known as polyester wrap, are commonly used to enhance the aesthetic appearance in residential upholstered furnitures.

Rxfibron HT PET is the first commercially released medical-grade, high-tenacity PET [10]. It is ideal for applications such as sutures and in endovascular, vascular surgery and orthopedics.

Commercially available angioplasty balloons are commonly made of PET, and catheters used can be of polyurethane, polyethene, styrene-butadiene, rubber or polystyrene.

Medically PET has several biomedical applications, such as in artificial ligaments and osteointegration. Biocompatible and non-absorbable artificial ligaments from modified PET show significant abrasion and mechanical robustness. Highly porous PET-based fibrous mats are essential for drug delivery, wound dressing, implants, and tissue engineering scaffolds. These mats are reported with desired mechanical properties.

PET-based stents and balloons-hybrid stents based on metallic core and outer PET mesh led to the discovery of the MGuard Stent. MGuard Stent is now the Gold standard in stents and has undergone many successful clinical trials. They are called mesh-covered stents. Smarter grafted PET-based stents have heparin conjugated PET-based stents for enhanced biocompatibility. Commercially available angioplasty balloons are commonly made of PET, and catheters can be of polyurethane, polyethene, styrene-butadiene rubber or polystyrene (Fig. **13**).

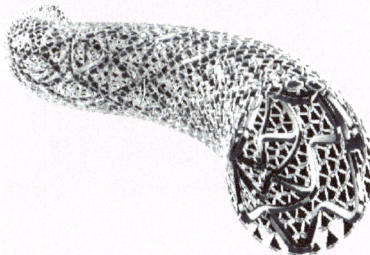

Fig. (13). MGuard Stent. Source: researchgate.net.

Shape-memory PET /polyethene blended thermo-responsive shape-memory polymers were synthesized using ethylene-butyl-acrylate-glycidyl methacrylate terpolymer. Shape memory PET was used as an implantable medical device.

Flexible electronic devices based on PET flexible, metal insulator, metal capacitors have been fabricated in medical electronics. These devices show excellent electrical stability and high mechanical flexibility.

Medical films with antimicrobial properties are essential for medical devices, sterile packaging and sanitization. There are several reports on polyetheneterepthalate-based compounds have been formed. These have cationic biocompatible materials like chitosan, polyethyleneimine, polyethylene glycol amine, *etc*.

Dacron is a trademark of DuPont for polyester fibre. These fibres have high tensile strength, high resistance to stretching in wet and dry conditions, and good resistance to degradation to bleach and abrasion.

This filament yarn is used in curtains, dress fabrics, high -pressure fire hoses and in men's apparel.In addition, the staple fibre is ideal for mixing with wool for clothing, knitted wear, and woven sportswear.

POLYOLEFINS

The word polyolefin means 'oil-like'. It comes from the oily or waxy texture of plastic types that include polyethene. It is a collective term used for the kind of plastics that includes low-density polyethene (LDPE), linear low-density polyethene (LLDPE), high-density polyethene (HDPE) and polypropylene (PP). Polyolefin raw materials are cost-effective. Their densities are the lowest of all plastics. As a result, polyolefin fibres are successfully recycled into new plastic products. Furthermore, recycling rates of polyolefin can increase if consumers separate their recyclables from the rest of their waste (Figs. **14** & **15**).

Fig. (14). Polyolefin straws.

Fig. (15). Transparent polyolefin.

Polyolefin is produced from carbon-rich substances such as coal, oil or natural gas. Polyethene, a substance very common to all polyolefins, is made up of many molecules of ethylene linked together to become a polyethene polymer.

LDPE-Low-density polyethene and linear low-density polyethene (LLDPE) are solid and rigid as used for general packaging and wrapping of consumer goods. They generally appear translucent or foamed. This LDPE is used as shrink films, carrier bags and heavy-duty refuse bags. In addition, LLDPE is used as stretch film, industrial packaging film, thin-walled containers and heavy-duty medium and small bags.

HDPE-High density polyethene packaging has excellent chemical resistance and therefore, can be used for bottles and drums for keeping chemicals. When handled, HDPE can also be made into opaque, translucent materials and identified by a crackling sound. Commercially they can be produced into crates and boxes, bottles for food products, detergents, cosmetics, industrial wrappings, carrier bags and drums for food, drinks and chemicals.

Polyolefin fibres can be produced as continuous filament yarn and monofilament. Among different types of polyolefin fibres, texturized continuous filament yarns are common. Some polyolefin pulp products are designed for use in paper and corrugated board production. Also these products can be used as fibrous fillers in concrete, and as a replacement for asbestos.

The largest end-use market is represented by nonwoven fabrics depending on their manufacture and characteristics and has many applications commercially. These uses include personal care, sanitary and medical applications, agriculture fabric, construction sheeting, industrial wipes, geotextiles, automotive material and filtration media. Historically, the primary end-use market of polyolefin fibres is in carpet and rugs. These fibres can be used to manufacture tufted and woven carpets, rugs and synthetic turf. Other uses include carpets, rugs, carpet backings, and face fibre. These fibres are also used in nonwoven in sanitary cover stock, filtration, medical application, home furnishings, geosynthetic fabrics, floor coverings, wiping clothes, industrial garments, construction, and durable papers.

Polypropylene is the predominant polyolefin fibre and has extensive use in textile applications. The physical properties of polypropylene and polyethene, especially tensile strength, abrasion resistance and inertness to most chemicals, even water, make them well-suited for functional applications though predominantly used in nonwoven in carpets and rugs. Still, these fibres have fundamental property limitations, such as poor dye ability and inability to maintain a crease or wrinkle, which has generally limited their growth potential in apparel and home fabric markets (Fig. **16**).

Fig. (16). Products made up of polypropylene.

The inherent stiffness of polypropylene (PP) material makes it ideal for thin-walled packaging items. Various kinds of films from polypropylene are used for different applications. They are different in the way they look and behave. These films are shiny, transparent and crackle when handled; therefore, they are ideally suited for lid closures and containers like tablet vials. The bulk of bottle caps and closures are made of polypropylene. Food packaging, including yogurt and margarine tubs, comprises of polypropylene. Buckets made up of polypropylene are used for detergents, peanut butter, paint and animal feed. Keeping stationery items, confectionery and even clothing, polypropylene containers and packaging can be used. Polypropylene films are also used to manufacture tapes for woven tape applications such as cement bags, bulk feed bags, *etc.* Non-woven textiles applications include personal hygiene products.

Polypropylene netting applications include safety fencing around horticulture building sites, to control plants' sunlight intensity. It is also utilized as a base to fill materials in road construction.

Until 2000, polyolefin fibres had been one of the fastest-growing segments of the synthetic fibre industry. Since 2000, the growth has slowed down due to an escalation in the price of polypropylene resins, but the market has recovered over the last couple of years. The demand for industrial applications such as ropes, nets and FBIC bagging has grown throughout all the regions. India is expected to have the fastest demand growth. As the population has increased and the standard of living has spurred the consumption of polyolefin fibres for textiles and in the non-woven sector for sanitary and technical applications like geotextiles, filtration purposes, *etc.* Slit film consumption for woven sacks and bags is also overgrowing as India produces them for domestic and export markets.

Using polyolefins blended with natural fibres, such as hemp based, as reinforcing filler in thermoplastic is a relatively new application. However, it has excellent potential to replace glass fibre products in the automotive industry [11]. Studies indicate that hemp-based natural fibre mat thermoplastics (NMT) are promising material in automotive applications requiring high specific stiffness.

High melt flow polyolefin materials manufacture optical fiber cable components such as buffer tubes, cores, filler rods and jackets. The polyolefin materials obtained from polymers of polyethene and polypropylene is a copolymer of propylene and ethylene or a tetra polymer that includes propylene and ethylene. This may be used to fabricate polyolefin optical fibre cable components with improved strength, compression resistance, reduced shrinkage, reduced process-induced orientation, and high crystallinity with improved processability. A fibre optic cable component manufactured from polyolefin is characterized by a higher Melt Flow Index (MFI). Thisis generally higher than that specified for 'extrusion grade' materials. By the manufacturers of polymeric materials, 'extrusion grade' materials are generally characterized by a low Melt Flow Index.

Optical fibre cables [12] are classified as loose tube optical fibre cables, which is generally filled with water-blocking compounds such as gel. These flexible buffer tubes arc stranded around a central member. In addition to buffer tubes, filler rods may be left around the prominent central member. These filler rods may be made of solid or cellular polymer. In a slotted core optical fibre cable, the optical fibres reside in channels or slots ,generally filled with a water-blocking gel.

The optical fibres reside in a central tube filled with a water blocking compound in the mono-tube cable. The buffer tube or core provides the primary structure, the thin optical fibers contained within. Buffer tubes or core is jacketed with an additional protective layer. Additionally, reinforcing yarns or threads, as well as water-blocking materials in the form of gels, hot melts, and may place water-swellable powders, yarn types, or corrugated armour between the jacket and the inner cable-layers.

Industrial applications [13] of polyolefin fibres have grown recently. However, their functional properties and cost determine the use of high-performance polyolefin fibres. Price performance ratios of polyolefin fibres, such as tenacity/price and modules/price, are very favourable when compared with polyester and polyamides. Recently polyolefin materials entered the technical textile markets due to enhanced fiber extrusion technology and advanced non-woven production processes such as melt-blown and hydro-entanglement technology.

The use of polyolefin fibres in medical textiles [14] is due to excellent chemical resistance and inertness, which are prerequisites for medical and surgical applications. These fibres can be used in non-implantable materials like wound dressing, bandages, plasters *etc.*, and in implantable materials like sutures, vascular grafts, artificial joints, surgery meshes, and artificial ligaments. In addition, healthcare and hygiene products are used as bedding, surgical gowns, wipes, diapers, *etc.*

It is used in the worldwide automotive marine and air transportation market [15], where it has replaced metals, rubber and glass with lighter material like this, saving fuel during transportation. It is mainly used in resin form for moulded parts such as door panels, armrests, instrument panel, sun visors, mirror housing, headliners, *etc.* Readily visible polyolefin are in the manufacture of carpets, interiors, in boat liners. Invisible ones are in tyre cords, airbags, and seat belts. These fibres resist chemicals and staining therefore, can be used for seat covers.

It is used in ropes, nettings, wines and cordages [16] in the fishing industry due to the inherent resistance of polyethene to chemical attacks, marine growth of microorganisms or other unwanted materials. Resistance to rotting, with excellent wet abrasion resistance with a long life period, retains its flexibility at extremely low temperatures that it even does not become rigid under freezing conditions.It does not shrink on immersion in water, on sturdy, smooth surfaces, does not cling to sand particles, makes the net easier to handle and saves time in shooting and hauling. It requires minimum maintenance.

Ropes made up of high-density polyethene monofilament yarns have many applications, from painters on dinghies to mooring ropes for tankers. In addition, polyolefin yarns are used in fishing gear including fishnets, trawls and lines, due to their high wet and dry strength, toughness, low weight, and resistance to weather, and chemical degradation, micro-organisms or commonly used degradative agents. Earlier, Nylon was used since the end of World War II. Still, polyester, polyethene and polyvinyl alcohol fibers have a place for manufacturing fishing gear in the world market.

POLYAMIDES

These fibres are derived from carbon-based molecules and are entirely synthetic. Nylon, aramids are types of polyamide.

NYLON

These fibres are made from a monomer called diamine acid extracted from crude oil. Diamine acid is forced to react with adipic acid to form a polymer known as nylon salt. DuPont Corporation originally developed this in the mid -1930s as an alternative to silk stockings, as it was more substantial than steel and entirely resistant to runs, thus changed the name from 'Nuron' to 'Nilon', later replaced 'l' with 'y' to **Nylon.**

The crystallized form of nylon salt was heated to form a molten substance. In DuPont Corporation, this substance was then extruded through a metal spinneret. Then it is loaded onto a type of spool called a bobbin. Fibres produced are stretched to increase their strength and elasticity. Then they are wound onto another spool. This process is known as drawing. The resultant fibres are then spun into textiles and other forms of threads.. This type of fibre is mixed with different materials woven into textile products or has other commercial applications.

During World War II, Nylon was commonly used as parachute material. Later a shortage of fabrics led many women to make dresses made up of parachute material. However, after a brief period, consumers quickly recognized that pure nylon was not well suited as a textile material because of its low breathability and damage prone when exposed to heat. Nylon was then blended with cotton, polyester or wool to give unique benefits such as silkiness and elasticity (Figs. **17** & **18**).

Fig. (17). Nylon bag.

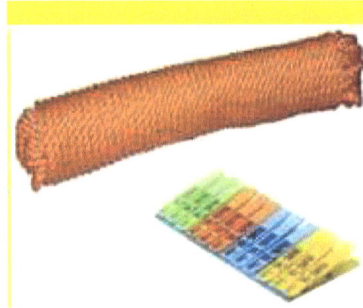

Fig. (18). Nylon rope and clips.

Nylon is a generic designation for a family of synthetic polymers based on aliphatic or semi-aromatic polyamides. Nylon is a thermoplastic silky material which can be melted and processed into fibres, films or shapes. Nylon-66 and nylon -6 are examples of condensation polymers made of polyamides. In most circles, nylon and polyamide fabric are synonymous and, this polymer is a type of polyamide fabric that has potential applications which are now significantly limited.

The term polyamide refers to a molecule that has repeating amide bonds. For example, silk and wool are polyamides but not synthetic; polyamide is synonymous with Nylon.

Nylon 6 & Nylon 66

Nylon6 & Nylon 66 are synthetic polymers or polyamides that are the essential fibres of this class. Nylon 6 fibres are made up of organic compounds which contain 6-carbon atoms.

Nylon6 is made from caprolactam. It is a type of monomer. Whereas Nylon 66 consists of two monomers. These monomers are adipoyl chloride and hexamethylenediamine. A strong chemical bond gives nylon 66 a significant crystalline structure. Consequently, this makes it stiffer and much better equipped to handle heat than nylon 6.

The greatest strength of Nylon 6 is its flexibility which replaces metals in products. It is generally used for making radiator grills, firearm components, stadium seats, bristles for cleaning brushes, plastic cutting boards, hammerheads, and circuit breakers . Due to its high tensile strength, nylon is also used to make rock climbing ropes.

Nylon 66 is more durable and has a higher melting point than nylon 6. Moreover, it can withstand heat and long-term wear and tear; therefore, it is a popular choice for products like battery moulds, conveyor belts, friction bearings in luggage, and in durable carpeting.

ARAMIDS

Aramids are the type of polyamide fibre that is commonly used in consumer and military applications but not in apparel. The most popular aramids are **Nomex** and **Kevlar**, which are widely used for clothing (Fig. **19**).

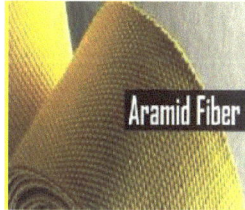

Fig. (19). Strip made of Aramid Fiber.

Aramids are raw materials used for body armour, bulletproof vests, firefighters' uniforms as protective wear, and firefighting gear. Aramid fibres are the superstars in the fibre world. Although colors are also available, the basic form appears bright golden in color. Aramid's name comes from two words 'aromatic and polyamide'. These fibres fibers have outstanding strength-to-weight ratio and heat-resistant character. Hence, these fibres play an essential role in the application as composites, automotive industry, in military application related fields.

During the development of aromatic polyamides, a lyotropic liquid crystalline aramid compound was discovered. This aramid compound was commercially known as **Kevlar,** which is an American company DuPont's brand, which was invented in early 1970 by a DuPont Chemist Stephanie Kwolekas as pre-aramid when she was looking for some light-weight, strong fibre to use for tyres. This enabled her to develop a novel spinning process which subsequently led to the commercialization of Kevlar (Fig. **20**).

Fig. (20). Kevlar fiber.

In the 1960s, before the invention of Kevlar, a meta-aramid fiber was produced by DuPont by a Scottish scientist Dr Wilfred Sweeny. The trade name of this was **Nomex** (Fig. **21**). Nomex was marketed in 1967. Meta-aramids are made in many other countries. Such as in Japan by Teijin under the trade name **Conex**. In Korea, Toray by the trade name **Arawin**, in China by the trade name **New Star** and **X-Fiber**. France also manufactures a variant of meta-aramid under the trade name **Kermel**.

Fig. (21). Fire fighter suit made of Nomex.

Both Kevlar (a para-aramid) and Nomex (a meta-aramid) have several variants with specific properties. Structurally, Kevlar is a kind of polyamide, where the amide groups are separated by phenyl rings attached to it at carbon one on one side and carbon four on the other amide group.

While Nomex is a polyamide with meta-phenyl groups in this amide, groups are attached to phenyl at carbon one and three positions. For Kevlar and Nomex, other meta- and para-aramid yarns can act as a replacement for less cost.

For Kevlar, Teijin has **Twaron** and **Technora**, **Heracron** by Kalon and **Alkex** by Hyosung is produced. For Nomex it is the most popular. **Teijin conex** under Teijin imparts similar results.

The characteristics of Kevlar fibres have excellent thermal resistance that does not catch fire or melt quickly. Thus, Kevlar is used to manufacture of protective clothing in air filtration units. It also acts as a substitute for asbestos.

NOMEX

Nomex is a meta-aramid marketed in 1967 and saved millions of lives, including aircraft pilots, racing car drivers, and firefighters, to name a few. This fibre has long-lasting thermal stability. It doesn't become brittle, soften or melt even if exposed to 300 degree centigrade. It is flame-resistant, self-extinguishing, and not affected by organic solvents. It is often used for electrical insulation.

Due to their outstanding properties, Aramid fibres have a wide variety of commercial uses, such as flame-resistant clothing, in military suits, helmets, – bulletproof wear and body armour.

When combined with carbon fibre, these fibres form composite materials. These are used for reinforced thermoplastic pipes, mechanical rubber reinforcement material and utilized mainly in tyres as **Sulfron**. Sulfron is also known as sulfur modified **Twaron**. Other commercial uses include jet engine enclosures, fibre reinforced concrete, loudspeaker diaphragm, sail clothes, boat hull materials and in optical fibre cable systems.

Today polyamide fibres represent about twelve per cent of global synthetic fibre production. Due to its softness and elasticity, it cannot effectively retain heat. However, these fibres can be woven into thin stockings. Other consumer goods made with polyamide are different automotive parts like toothbrushes, hair combs, firearm components and suitable packaging materials.

Environmental Impacts- Nylon and other polyamides fibres or their manufacturing process damage ecosystems. Water that is used to cool polyamides introduces contaminants into the ecosystem. . For example, one of the chemicals used to create nylon is adipic acid, which releases nitrous oxide into the environment, 300 times worse than CO_2. Fabrics made from polyamides are not biodegradable, which means they are permanent pollutants to the environment.

POLYVINYL FIBERS

Polyvinyl fibres are a type of fibre in which fibre forming substance is any long-chain synthetic polymer composed of at least eighty-five per cent by weight of vinyl chloride units. They are also known as polyvinyl chloride fibres, **Vinyon** fibers or **Chlorofibres**. These are used to make elastomeric fabrics (Fig. **22**).

Fig. (22). Elasticity of chlorofibres. Source: textilelearner.net.

Another type of fiber is **PVA fibre**, polyvinyl alcohol, a high-performance reinforcement fibre generally used for concrete and mortar. These PVA fibres uniquely designed to form molecular bonds with mortar and concrete.

Vinyon

Polyvinyl chloride fibres or **Vinyon** have eighty five per cent by weight of vinyl chloride units (-CH2-CHCl-). They are copolymers of vinyl chloride and vinyl acrylate or ethylene-vinyl acetate. Vinyon was initially a trademark of Union Carbide for polyvinyl chloride fibres. These are copolymerized with acrylonitrile called Vinyon N. By the 1950s, FTC received Vinyon as the general term for PVC fibres (Fig. **23**).

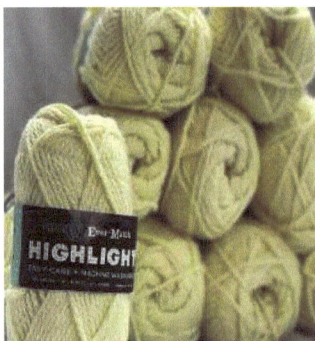

Fig. (23). Vinyon yarn, Source: Etsy.com.

Polyvinyl chloride (PVC) fibres was first invented in Germany in 1931. Commercially Vinyon fibre was produced by American viscose in 1939, now FMC corporation in United States is commercially a major manufacturer of Teijin and Denka vinyon fibres.

Vinyon is durable, has good retention properties, is resistant to weathering, and is less heat resistant than nylon and polyester. Pure PVC fibres are mainly used in outdoor fabrics like rain gear, traps, and fishing nets. When exposed to oxygen and excessive heat, they start to degrade and polymer properties begin to deteriorate; therefore, these fibres are seldom used for apparel that must withstand ironing. Vinyon fibres are impervious to chemicals, microbes and bugs. They are used as fire safe fibres in kids' clothing, as well as cover draperies and coverings. These fibers are additionally utilised as a part of angling nets, twins, and felts, as industrial fabric for blankets, canopies and open-air furniture.

POLYVINYLIDENE FIBERS

This long chain synthetic fibre consists of at least eighty percent by weight of vinylidene chloride.

Saran Fibers

These are materials with a greasy dark green film, called **Econite** and then **Saran**. Saran fibres come in monofilament, multifilament-twists. It is also available in thermochromic, which means 'colour changing properties and luminescent, which glows in the dark. Saran is a trading name for some polymers made from vinylidene chloride (PVDC). Ralph Wiley accidently discovered polyvinylidene chloride in 1933.

Dow Chemicals in 1940 under the trademark **Saran,** commonly used as thin films and fibres.

This transparent, very smooth fibre is resistant to many chemicals, including most organic solvents. Saran fibre has excellent abrasion resistance. Resistance to sunlight has self-extinguishing properties, resistance to flame, with low moisture retaining properties, makes it all weather resistant, which makes this fibre an excellent choice for underwater and outdoor exposure applications. It is used for home and automobile upholstery, sport shoes and bags where it needs more demanding application. Saran fibres are generally used for high performance applications. The most important uses are for inherently non-flammable uniforms and for filters where it resists aggressive chemicals (Figs. **24** & **25**).

Fig. (24). Saran fibres for dolls hair. Source:feltmangnate.com.

Fig. (25). Clear wrap made of Saran.

Electro-spun saran nanofibres [17] have a lot of potential in the protective clothing sector. Therefore, significant steps have taken to develop protective clothing with high breathability and elasticity, focusing on exceptionally lightweight fabric with multifunctional characteristics.

In households, saran fibres are used for cleansing cloths and filters. Industrial applications include screens, artificial turfs, and waste-water treatment materials.

Miscellaneous applications are also used for high-quality doll hair for its shine and softness.

POLYURETHANE FIBRES

Polyurethane fibres are sometimes called **Spandex** or **Elastane.** Fibres forming substance consists of approximately eighty five per cent segmented polyurethane, which is reacted with diisocyanate.

These fibres have rubber-like qualities and are more durable than rubber, which can be stretch more than five hundred per cent without breaking it. It is abrasion-resistant, wrinkle-resistant and excellent elastic fibres. It burns slowly in flame. Spandex has excellent resistance to mildew and does not shrink when exposed to water. Thermoplastic polyurethane can be moulded and shaped. Most of these elastomers can be spun into fibres. Spandex fibres can be used to make elastic textiles such as socks, tops, brasiers, support hoses, waterproof clothing, swimsuits and other athletic apparel.

The most popular brand names of polyurethane yarns are **Lycra, Dorlastan, Glospan** and **Cleerspan**, **Roica and Linel**. Owing to high elongation properties, low density transparency and other physical and chemical features, polyurethane yarn has found wide applications in the textile industry, replacing rubber yarn, which was once used in the production of bathing suits and later in corsets and hosiery. In addition, polyurethane yarns are used in practically all products used in knitting techniques [18]. These yarns with a multifibre structure are used widely in the knitting industry. The threads are produced as matt, transparent or translucent, depending on their intended purpose. In addition, these yarns exhibit high resistance to acid, alkali, oils and fats and chlorine and are thus resistant to yellowing.

Lycra

It is the most popular form of spandex [19] that is body hugging and form-fitting, commonly used for sheer undergarments for men and women. Most undergarments that hug close to the skin contain at least some amount of this fibre (Fig. **26**).

Fig. (26). Lycra pants.

Lycra is a brand name for **Elastane**. Despite having different term, **Lycra**, **Spandex** and **Elastane** are all the same material. Fabrics made up of these can stretch 5-8 times their usual size (Fig. **27**).

Fig. (27). Spandex, Source: sewport.com.

Any fibres can mix with Lycra fibre. A minimal amount of lycra fibre can transform the performance of a yarn in which it is mixed. There are various ways of integrating lycra fibre with other threads. Depending on the weaving or knitting methods and varieties of natural or artificial fibres blended, many applications of lycra are possible. It can be used as innerwear such as pantyhose, hosiery, swimwear, cycling jersey and shorts, exercise apparel, leotards *etc.* In fashion clothing, materials such as elastic tapes and cords are used in hygiene products such as disposable diapers. Diapers [20] are the emerging area of nonwoven applications in personal hygiene, such as baby and adult diapers, feminine hygiene products and adult incontinence products. Lycra can be used in compression garments as foundation garments, motion capture suits, *etc.* It is also used as a surgical hose.

CONCLUSION

In recent decades, most research globally has been concerned with different methods of incorporating synthetic fibre with natural fibres to modify their properties, decrease their cost and increase their strength. Therefore, tensile and impact test were conducted on different combinations of fiber composites.

Since synthetic fibres are much cheaper and can be an alternative to natural fibre, the popularity of synthetic fibres has grown a lot. For example, the demand for polyester fibre has risen sky-high, and it is now the single most used fibre for textiles, overtaking cotton. However, polyester clothes look very neat, but once it is worn there might be breathability issues, and during sweating, it turns sticky.

Synthetic fibres have many advantages over natural fibres, but it has disadvantages too, like it burns more readily than natural fibres and is, therefore, more prone to heat damage. It melts relatively quickly. Hot washings damages the fibre. Rubbing these fibres generates an electrostatic charge. It is not skin friendly, so it is uncomfortable for extended wear; hence it can be allergic to some people.

It is a common misconception that manufacturing of consumer goods from natural fibres doesn't hamper the environment, whilst the synthetic ones damage the environment and the ecosystem. But this isn't true. For example, cotton and polyester have similar environmental impacts due to having similar manufacturing styles with chemicals additives like bleaches and detergents; they can be proven toxic to the environment.

It is a matter of growing concern that factory workers get exposed to synthetic fibers. Inhaling tiny synthetic fibres or its dust leads to various broncho-pulmonary diseases [21] like asthma, extrinsic allergic alveolitis, chronic bronchitis, spontaneous pneumothorax and chronic pneumonia.

CONSENT FOR PUBLICATION

Not applicable.

CONFLICT OF INTEREST

The authors declare no conflict of interest, financial or otherwise.

ACKNOWLEDGEMENT

Declared none.

REFERENCES

[1] Kadolph S, Langford A. Textiles. 9th ed., New York: Prentice-Hall 2002.

[2] Periyasamy AP, Durai B, Thangamani KF. An overview of processing and application of lyocell. Textile Review 2011.

[3] Jiang X, Bai Y, Chen X, Liu W. A review on raw materials, commercial production and properties of lyocell fiber. Journal of Bioresources and Bioproducts 2020; 5(1): 16-25.
[http://dx.doi.org/10.1016/j.jobab.2020.03.002]

[4] Eva B. Lyocell is the new generation of regenerated cellulose. Acta Polytech Hung 2008; 5: 11-8.

[5] Wang L J. Y.N., Liu, D. F.and Li, Z.J. (2017). Status and development of lyocell fiber at home and abroad. *J Textile Res* 38: 164-170.

[6] Woodings C. Regenerated Cellulose Fibers. Cambridge: Woodhead publishing Limited 2001.
[http://dx.doi.org/10.1533/9781855737587]

[7] Militky, J. The chemistry, manufacture and tensile behavior of polyester fibers. Handbook of Textile properties of Textile and technical Fibers. Woodhead Publishing Series of Textiles 2009: 223-314.

[8] Fangueiro R, Pereira G, Araujo MDe. Applications of polyesters and polyamides in civil engineering. Polyesters and Polyamides. Woodhead Publishing Series in Textiles 2008; 542-92.
[http://dx.doi.org/10.1533/9781845694609.3.542]

[9] Nazare S. S. and Davis, R. D. 2013.Flame retardancy testing and regulation of soft furnishings, *Handbook of Fire-Resistant Textiles.*. Woodhead Publishing Series of Textiles. 456-498 2013.
[http://dx.doi.org/10.1533/9780857098931.3.456]

[10] Farah S, Kunduru K R, Basu A, Domb AJ. Molecular weight determination of polyethylene terephthalates. In: Visakh PM, Liang M (Eds). Poly(ethylene terephthalate) based blends, Composites and nanocomposites, William Andrew, Elsevier 2015: 143-165.
[http://dx.doi.org/10.1016/B978-0-323-31306-3.00008-7]

[11] Pervaiz M, Sain MM. Sheet molded polyolefin natural fiber composites for automotive applications. Macromol Mater Eng 2003; 288(7): 553-7.
[http://dx.doi.org/10.1002/mame.200350002]

[12] Risch BG, Holder JD. 1999. Polyolefin materials suitable for optical fiber cable components United State patent, Risch et al. Patent Number 5911023.Date of Patent June 08.

[13] Ugbolue SCO. Polyolefin fibers: Structure, properties and industrial applications. 2nd ed., Woodhead Publishing: Elsevier 2017.

[14] Horrocks AR, Anand SC. Handbook of Technical Textiles, Technical Textile Applications. Cambridge, UK: Woodhead Publishing Ltd., 2016.

[15]   Ogaki Y, Berman L. Automotive Cabin Filters. Technical Textiles International 1996; 4(10): 16.

[16]   Foster GP. New high-tech fibers bring high strength–to–weight resolution for ropes. Proceedings of Techtextil North America Symposium. Atlanta, GA. 2000.

[17]   Gorji M, Bagherzadeh R, Fashandi H. Electro spun nanofibers in protective clothing.Electro Spun Nanofibers. Woodhead Publishing Series in Textiles 2017; pp. 571-98.
[http://dx.doi.org/10.1016/B978-0-08-100907-9.00021-0]

[18]   Mielnicka E. Types and suitability of yarns for knitting. In: Au KF (Ed.) Advances in Knitting Technology. Woodhead Publishing Series in Textile 2011; 03-36.

[19]   Langley KD, Y.K. Kim. 2006. Manufacturing nonwovens and other products using recycled fibers containing spandex. In: Wang Y. (Ed.) Recycling in Textiles.Woodhead Publishing Series in Textiles 2006; 137-64.

[20]   Ajmeri J. R.and C.J. Ajmeri, C.J. 2016. Developments in the use of nonwovens for disposable hygiene products. In: Kellie G. Advances in Technical Nonwovens. Woodhead Publishing Series in Textiles 2016; 473-96.

[21]   Pimentel JC, Avila R, Lourenço AG. Respiratory disease caused by synthetic fibres: a new occupational disease. Thorax 1975; 30(2): 204-19.
[http://dx.doi.org/10.1136/thx.30.2.204] [PMID: 1179318]

CHAPTER 5

Investigation of Substrate-effect on BaF_2 Thin Films: A Study of Fractal Nature

Pradip Kumar Priya[1,*], Ram Pratap Yadav[2], Hari Pratap Bhasker[3], Anil Kumar[4] and Kusum Lata Pandey[1]

[1] *Department of Physics, Ewing Christian College, University of Allahabad, Prayagraj-211003, India*

[2] *Department of Physics, Deen Dayal Upadhyay Govt. P.G. College, Saidabad, Prayagraj-221508, India*

[3] *Department of Physics, Chaudhary Mahadeo Prasad Degree College, University of Allahabad, Prayagraj- 211002, India.*

[4] *Department of Physics & Electronics, Dr. Ram Manohar Lohia Avadh University, Ayodhya-224 001, India*

Abstract: BaF_2 thin films of thickness 20 nm are prepared using the electron beam evaporation technique (at room temperature) on glass, silicon (Si) as well as aluminum (Al) substrate, respectively. These substrates play a crucial role in the evolution of thin film surface morphology. The thin films grown far from equilibrium have self-affine nature which is reminiscent of fractal behaviour. The surface morphology of films is recorded by atomic force microscopy (AFM). Scaling law analysis is performed on AFM images to confirm that the thin film surfaces under investigation have self-affine nature. The concept of fractal geometry is applied to explore-how different substrates affect the surface morphology of films. The fractal dimension of horizontal as well as vertical sections of AFM images are extracted by applying Higuchi's algorithm. Value of Hurst exponent (H) for each sample is estimated from fractal dimension. It is found to be greater than 0.5 for Al as well as glass substrates, indicating that the height fluctuations at neighboring pixels are correlated positively. However, for Si substrate, its value is less than 0.5 which suggests that the height fluctuations at neighboring pixels are not positively correlated.

Keywords: BaF_2 thin film, Atomic force microscopy (AFM), Self-affine, Hurst exponent, fractal dimension.

* **Corresponding author Pradip Kumar Priya:** Department of Physics, Ewing Christian College, University of Allahabad, Prayagraj-211003, India; E-mail: aurampratap@gmail.com, hpb.bhu@gmail.com

Arti Srivastava, Mridula Tripathi, Kalpana Awasthi and Subhash Banerjee (Eds.)
All rights reserved-© 2023 Bentham Science Publishers

INTRODUCTION

A famous quote about fractal by Benoit Mandelbrot *"Clouds are not spheres, mountains are not cones, coastlines are not circles, and bark is not smooth, nordoes lightning travel in a straight line"* [1]. There are many examples in nature such as mountains, rivers, coastline, leaves, air bubbles, cloud, *etc.*, which show the fractal characteristics. Modeling of such natural fractal behavior is impossible through a classical technique which is based on assumptions of smoothness. Over the time, many simple mathematical models were developed to deal with the fractal behavior. Some of them are Cantor set, Sierpinski Gasket, Sierpinski Carpet and Koch curve. Fractal can be defined as a set, for which Hausdorff Besicovitch dimension D (called fractal dimension) exceeds its topological dimension D_T [2]. In general, fractal objects remain invariant under certain transformations of scale.

Thin film exhibits unique physical and chemical properties as compared to its bulk counterparts [3]. The process of thin film deposition is highly crucial for fabrication of advance devices such as microprocessors, memory storage devices, photovoltaic devices, supercapacitors, *etc.* Surface morphology is also important in various biomedical applications namely biosensors, DNA/RNA sensing, biomedical engineering, *etc* [4 - 6]. Surface morphology plays an important role in optical [7], mechanical [8], magnetic [9], electrical [10], and tribological [11] based devices. This requires a deep understanding about the evolution of surface morphology. Interestingly, the surface morphology of the films exhibits a fractal nature. Therefore, the precise control over the growth of good quality thin films is necessary for its applications in highly efficient devices. The substrate plays a crucial role in governing the properties of thin films. It is found that the different growth mechanisms are governed by different morphologies.

Barium fluoride (BaF_2) is a wide band gap material with a fluorite cubic phase structure and shows insulating properties [12 - 14]. BaF_2 can be used in various applications like CO_2 laser, low index optical coating, high stopping power and very fast scintillators, *etc.* It has also been employed in fabrication of single-layer antireflection coatings for NaCl optical components [14 - 16]. It was also used as a molten bath component in aluminium raffination and welding rod coatings and welding powders [16].

Classically the surface morphology of films is measured through average roughness and interface width [17]. It is worthy to mention that it cannot characterize the local roughness behavior which depends on the resolution of the measuring devices. The precise information about local roughness can be obtained through a parameter known as roughness exponent. This parameter governs the scaling of local surface height fluctuations with measurement of resolution. It is important to emphasize here that the growth of thin film surface is not a smooth process rather it is a well stochastic process. It means this behaviour is not linear at small scale and it shows fractal nature. There is a global need to characterize the complex or irregular structure of a surface using a scale invariant technique. The fractal analysis is the one which fits to provide such alternative by measuring fractal dimensions of the samples [18]. Fractal dimension can be determined indirectly by different techniques such as power spectral density (PSD) method, detrended fluctuation analysis (DFA), height-height correlation, *etc* [19 - 24].

An efficient algorithm was developed by Higuchi to compute the fractal dimension of a time series. This algorithm has been employed in various fields of science and technology. The fractal analysis can be used to study the brain dynamics through electroencephalographic (EEG) signals. The Earth's geomagnetic activity was also studied. Yadav *et al.* calculated the grain size, fractal dimension (D_f) and Hurst exponent (H) for different films of varying thickness as well as deposition angle [25]. Their study suggests that with increase of thickness of the film, there is a decrease in fractal dimension while Hurst exponent increases. This reveals that complexity decreases when the thickness increases. Talu *et al.* used the similar technique to establish a strong correlation between electrical resistivity and fractal dimension for Ag/Cu thin film [26]. Their finding suggests that the electrical resistance decreases sharply as the fractal dimension decreases [26].

In this chapter, we present detailed surface morphological studies using AFM images of BaF_2 thin films deposited by the electron beam evaporation technique on three different substrates. The Higuchi algorithm was applied to the digitized AFM data for estimating fractal dimensions of the horizontal as well as vertical sections of the film surfaces.

EXPERIMENTAL DETAILS

The evaporation technique was used for the deposition of BaF_2 thin films on glass, silicon (Si) and aluminum (Al) substrates at room temperature in vacuum [12]. The vacuum of the chamber was kept at $\sim 10^{-6}$ mbar. Standard cleaning process was performed for the cleaning of the substrates [12]. During this process, the rate of deposition was constant at $\sim 0.5 \times 10^{-9}$ meter/sec. The thicknesses of the films were measured *in-situ* using a quartz crystal monitor. The thickness of each sample was estimated to be 10^{-9} meter. The surface morphologies were recorded using atomic force microscopy (AFM). Further, the AFM images of size 2.0 μm×2.0 μm were digitized into 512×512 pixels [12].

METHODS

Classical Technique

Deposition of the film is a highly complex phenomenon and randomness is inherent in it. Due to this, roughness evolves in films. The mean deviation () and root mean square roughness (w) are the key parameters to study the surface morphology of films. Roughness, defined by root mean square (rms) deviation, is a manifestation of the randomness. It characterizes a global property of the film surface. If y(i,j) denotes the surface height at the pixel (i,j),

$$R_a = \left\langle \left| y(i,j) - \left\langle y(i,j) \right\rangle \right| \right\rangle \tag{1}$$

$$w = \sqrt{\left\langle [y(i,j) - \left\langle y(i,j) \right\rangle]^2 \right\rangle} \tag{2}$$

where <y(i,j)> represents the mean value of the heights over a surface of side L and is defined by

$$\left\langle y(i,j) \right\rangle = \frac{1}{L} \sum_{i=1}^{L} \sum_{j=1}^{L} y(i,j) \tag{3}$$

Here, it is important to mention that andw are not able to give any information about the lateral surface correlation which is an important characteristic of surface but measure the vertical properties of the surface. The correlation properties of the surface can be characterized using the autocorrelation function, height-height correlation function, and correlation length (ξ). The auto-correlation function is defined by:

$$A(r) = \left\langle h(x+r)h(x) \right\rangle_x \tag{4}$$

where h=y−<y> measures the deviation in height from the mean value. In general, for r>> ξ, autocorrelation function is not considerable.

The height-height correlation function is defined by [12, 22].

$$H(r) = \left\langle \left(h(x+r) - h(x) \right)^2 \right\rangle_x \tag{5}$$

Typically,

$$H(r) = \begin{cases} 2w^2 & \text{for } r \gg \xi \\ r^{2\alpha} & \text{for } r \ll \xi \end{cases} \tag{6}$$

For a smooth surface, the difference in surface heights at two points is proportional to the distance r while, for rough surfaces [12, 22],

$$F(r) \equiv \left\langle \left| h(x+r) - h(x) \right| \right\rangle_x \sim \left| r \right|^\alpha \tag{7}$$

where 0<α<1. When the lateral distance is scaled by a positive factorε then,

$$F(r) \equiv \left\langle \left| h(x+r) - h(x) \right| \right\rangle_x = \left\langle \left| h(\varepsilon x + \varepsilon r) - h(\varepsilon x) \right| \right\rangle_x \sim \left| \varepsilon r \right|^\alpha \sim \varepsilon^\alpha F(r) \tag{8}$$

This shows that the average height separation will get scaled by a factor ε^α. The exponent α is known as roughness exponent which is unity (α=1) for smooth surface. As surface becomes more and more rough, the value of αdecreases.

In the case of thin film surfaces, α measure the surfaces at short scale only. The fluctuations get saturated on a large scale and in this case, the interface width characterizes the surface roughness.

A random function X(x) is defined to be self-affine with scaling exponent H>0 if for anyb>0,X(bx)and bHX(x)have the same probability distribution. Such a function satisfies the following properties [12].

$$\left\langle \left| X(x+br) - X(x) \right|^2 \right\rangle_x = b^{2H} \left\langle \left| X(x+r) - X(x) \right| \right\rangle_x \tag{9}$$

$$\left\langle \left| X(x+r) - X(x) \right| \right\rangle_x \sim r^H \tag{10}$$

$$\left\langle \left| X(x+r) - X(x) \right|^2 \right\rangle_x \sim r^{2H} \tag{11}$$

Thus, the surface height function h(x) is a self-affine random function.

Brownian Motion

A discretized approximation to a Brownian motion traces yields an appropriate example of a random self-affine function [25]. A Brownian motion trace at discrete times $t_i = i\Delta t$, with ian integer, is calculated by summing up a series of random steps which can be described by the random function $B(i) = \sum_{j=1}^{i} R(j)$ where R(j) equals 1 or -1 with equal probability. Note that each R(j) is independent and does not depend on its preceding values. Thus, $\langle R(i+j)R(i+k)\rangle_i = \delta_{jk}$. The function B satisfies

$$\left\langle \{B(i+r) - B(i)\}^2 \right\rangle_i = \left\langle \left\{ \sum_{j=1}^{r} R(i+j) \right\}^2 \right\rangle_i = \sum_{j,k=1}^{r} \langle R(i+j)R(i+k)\rangle_i = \sum_{j,k=1}^{r} \delta_{jk} = r \tag{12}$$

Comparing eqn. (12) with eqn. (11) we can say that B is a random self-affine function with scaling exponent 0.5. One can verify that $B(bx)$ and $b^{0.5}B(x)$ have the same probability distribution for b>0. That is, when B is plotted and a portion of this plot, say one-fourth the size on the horizontal axis is magnified four times on the horizontal axis and two times on the vertical axis (Fig. **1**), the plot so obtained will have the same histogram as that of the original plot.

Fractional Brownian motion (fBm) is a generalization of the Brownian motion denoted byBH(t), that allows the scaling exponent to take any value in the range 0<H<1 so that

$$\left\langle \{B(i+r) - B(i)\}^2 \right\rangle_i = |r|^H \tag{13}$$

where H is the Hurst exponent. For Brownian motion when H=0.5, the change at any time step does not depend on the change in previous steps. However, for any other values of H, change in time step is correlated with a change in the previous step. For H>0.5, the correlation is positive which implies that the change at any time step is to be in the same direction as that of the previous time step. In other words, fluctuations grow in only one direction and this property is called 'persistence'. For H<0.5, the correlation among the successive time steps is negative and the change in it is in opposite direction to that of the previous step *i.e.* fluctuations get reverse quickly, known as 'anti-persistence'.

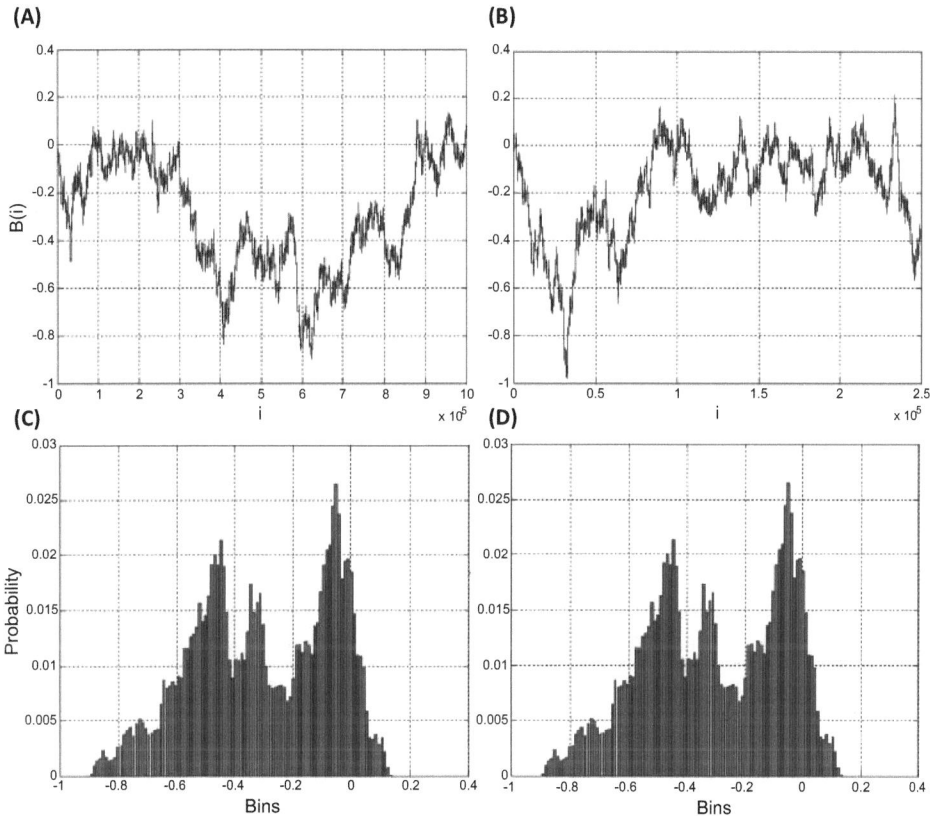

Fig. (1). (a) Position of a Brownian particle in 1-dimension after time *t*. **(b)** is obtained by considering a portion of the curve in **(a)**. Histogram plots **(c)** [before scaling] and **(d)** [after scaling] of Brownian curve shown in plots (a) and (b), respectively. Reprinted from "Effect of angle of deposition on the Fractal properties of ZnO thin film surface" by R. P. Yadav, 2017, Applied Surface Science, vol. 416, pp. 51-58. Copyright (2020) by Elsevier.

Here, it is needless to mention that the roughness exponent α is analogous to the Hurst exponent H for fractal Brownian motion. For smooth surface (α=1), the fractal dimension is 2. When the roughness exponent decreases the surface is no longer smooth.

Higuchi Algorithm

Higuchi proposed an algorithm which can be employed to evaluate the fractal dimension of vertical/horizontal section of a surface through the discretized height of an AFM data [21, 27]. This algorithm is capable of extracting the fractal dimension from a time series for the Brownian motion. This algorithm has widely been used to extract fractal dimension from observed data in different disciplines. The Higuchi algorithm is briefly described here [21, 27]:

The height at each pixel of a horizontal section of the surface is denoted by $Y(1)$, $Y(2)$.................$Y(N)$. The algorithm constructs k subsequences as

$$Y_m^k : Y(m).Y(m+k).Y(m+2k)........Y(m+pk), \tag{14}$$

where $m = 1,2,3.............. k$, and $p = \mathrm{int}\left[\dfrac{(N-m)}{k}\right]$

A measure $L_m(k)$ of the sequence Y_m^k is then defined by

$$L_m(k) = \frac{\left\{\left(\displaystyle\sum_{i=1}^{p} | Y(m+ik) - Y(m+(i-1).k) |\right)\dfrac{N-1}{pk}\right\}}{k} \tag{15}$$

where N denotes the complete set of the sample points and $\dfrac{N-1}{pk}$ is a normalization constant which takes into account the residual pixels not covered by the subsequence. The mean value of length is defined as,

$$L(k) = \frac{\displaystyle\sum_{m=1}^{k} L_m(k)}{k} \tag{16}$$

$L(k)$ scales with k according to the power law as given below

$$L(k) \sim k^{-D_f} \tag{17}$$

where D_f is called the Higuchi fractal dimension. The slope of log $L(k)$ *versus* log k directly provides the fractal dimension D_f where $L(k)$ is the average number of boxes that covers the $(i, Y(i))$ graph. Hurst exponent can be obtained from the fractal dimension using [28],

$$H = 2 - D_f \tag{18}$$

RESULT AND DISCUSSION

Fig. (**2**) shows the AFM image for BaF₂ films grown on Glass, Si, and Al substrates, respectively. It is clear from the the Fig. that the surface morphologies for different samples are distinctively different which indicat that substrates strongly affect the surface properties of the films.

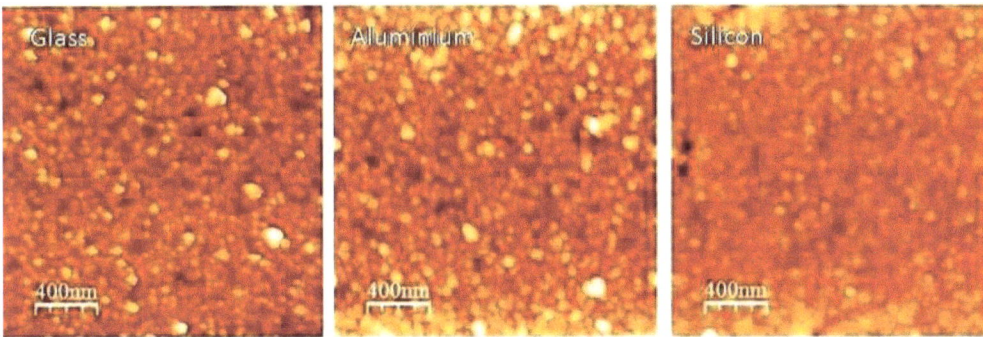

Fig. (2). The AFM images of BaF₂ thin film surface deposited on the Glass, Si and Al substrates. Reprinted from "Substrate effect on the evolution of surface morphology of BaF₂ thin films: A study based on fractal concepts" by Kavyashree, 2019, Applied Surface Science, vol. 466, pp. 780-786. Copyright (2020) by Elsevier.

In order to get detailed information about the samples, various surface parameters are calculated. The average roughness (R_a) was calculated using equations (1) and (2). Its values for different samples are given in Table **1**.

Fig. (**3**) shows the plot of $L(k)$ as a function of k at double logarithmic scale. The fractal dimension is estimated for each of 512 sections of AFM images in the fast scan direction (rows) and perpendicular direction. The calculated values of fractal dimension are given in Table **1**.

Table 1. The parametric values of average roughness (R_a), interface width (w), crystallite sizes (d), Fractal dimension (D_f) and Hurst exponent (H). Reprinted from "Substrate effect on the evolution of surface morphology of BaF$_2$ thin films: A study based on fractal concepts" by Kavyashree, 2019, Applied Surface Science, vol. 466, pp. 780-786. Copyright (2020) by Elsevier.

Substrate	R_a (nm)	w (nm)	d (nm)	Rows		Columns	
				D_f	H	D_f	H
Glass	3.39	4.02	15.5 ± 1.0	1.32 ± 0.03	0.68 ± 0.03	1.31 ± 0.04	0.69 ± 0.04
Si	2.41	3.02	12.4 ± 1.2	1.61 ± 0.04	0.39 ± 0.04	1.67 ± 0.04	0.33 ± 0.04
Al	4.86	5.68	15.3 ± 1.2	1.40 ± 0.04	0.61 ± 0.04	1.31 ± 0.04	0.69 ± 0.04

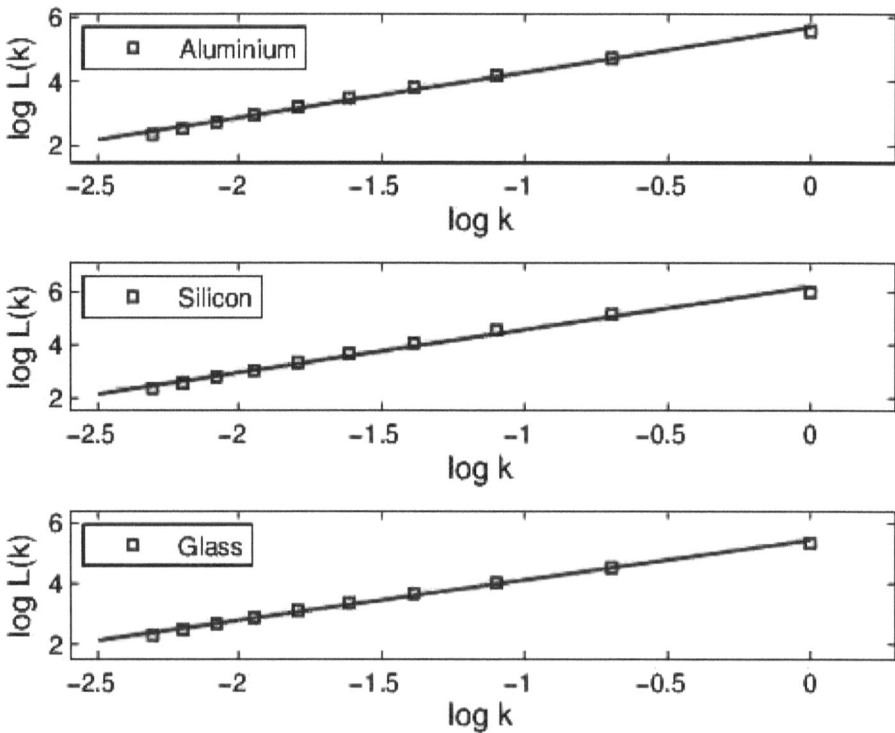

Fig. (3). Log L (k) as function of log k. The solid lines are best-fit curve. Reprinted from "Substrate effect on the evolution of surface morphology of BaF$_2$ thin films: A study based on fractal concepts" by Kavyashree, 2019, Applied Surface Science, Vol. 466, pp. 780-786. Copyright (2020) by Elsevier.

The slope of the plot between log $L(k)$ *vs.* log k directly provides the fractal dimension D_f of the samples.

The calculated values of the fractal dimension for rows and columns of each film are shown in Figs. **4(a)** and **(b)**.

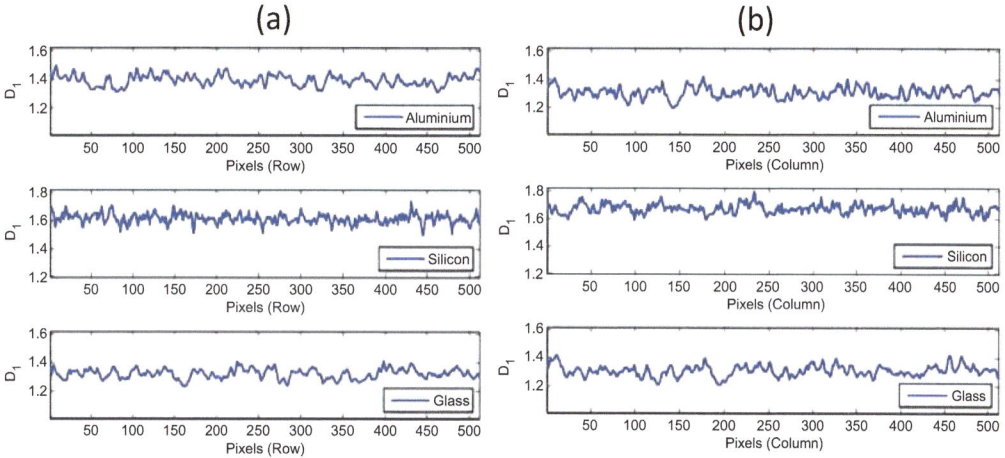

Fig. (4). (a) Fractal dimension (D$_f$) value for row pixels for the BaF$_2$ thin film on the Glass, Si and Al substrates. **(b)** Fractal dimension (D$_f$) value for column pixels for the BaF$_2$ thin film on the Glass, Si and Al substrates. Reprinted from "Substrate effect on the evolution of surface morphology of BaF$_2$ thin films: A study based on fractal concepts" by Kavyashree, 2019, Applied Surface Science, vol. 466, pp. 780-786. Copyright (2020) by Elsevier.

For each substrate, the fractal dimension for rows and columns are given in Table **1**. The fractal dimension for rows and columns of BaF$_2$ thin films deposited on glass substrate are estimated to be ~1.31 and ~1.32, respectively. These values for rows as well as columns are almost equal suggesting that surface growth is isotropic [29]. This isotropic growth may be due to the amorphous nature of the glass substrate. Growth on glass substrate is seen to be more isotropic than to that on Silicon as well as Aluminum substrates. It is important to mention that the higher fractal dimension suggests that the surface morphology is more irregular whereas the lower fractal dimension indicates smoother surface [30].

The estimated values of Hurst exponent (*H*) for different substrates as a function of rows and columns are shown in Figs. **5(a)** and **5(b)**, respectively.

For each substrate, the estimated values of *H* are also given in Table **1**. The beauty of Hurst exponent is that it provides the information of surface roughness or complexity of surface at small scales and its value lies between zero and one [2, 20]. Interestingly, when *H* = 0.5 it suggest that the height fluctuations at any pixel does not depend on the fluctuations in neighboring pixels [2, 20, 31]. For *H* > 0.5, the surface heights at neighboring pixels are positively correlated while for *H* < 0.5 the surface heights at neighboring pixels are negatively correlated.

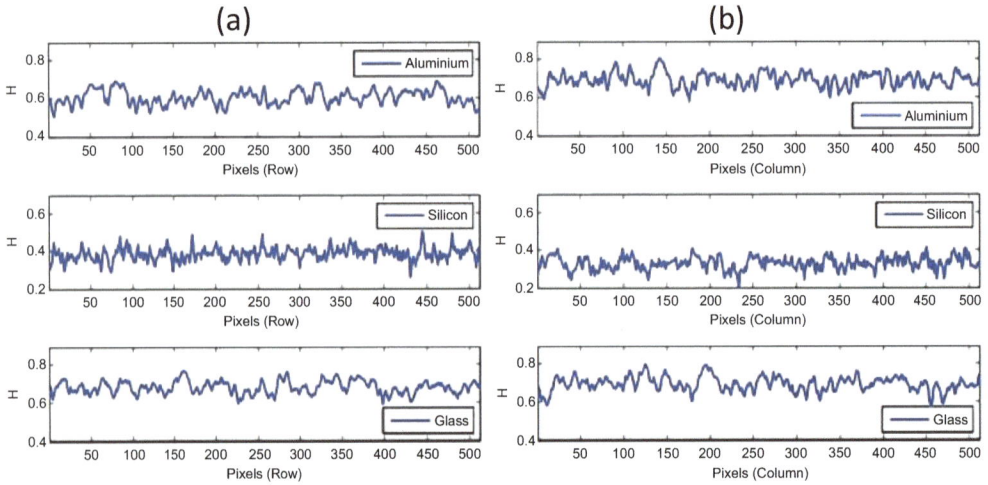

Fig. (5). (a) Hurst exponent (H) value for row pixels for the BaF_2 thin film onto the Glass, Silicon and Aluminum substrates, respectively. **(b)** Hurst exponent (H) value for column pixels for the BaF_2 thin film on the Glass, Si and Al substrates, respectively. Reprinted from "Substrate effect on the evolution of surface morphology of BaF_2 thin films: A study based on fractal concepts" by Kavyashree, 2019, Applied Surface Science, vol. 466, pp. 780-786. Copyright (2020) by Elsevier.

The Hurst exponent is found to be greater than 0.5 for both Aluminum and glass substrates. This suggests that there exist the 'persistence' during the deposition of films. While, for Silicon substrate, Hurst exponent comes out to be less than 0.5, indicating 'anti-persistence'. Therefore, for Aluminum and glass substrates, the height fluctuations in neighboring pixels are more likely to be in the same direction while that of opposite direction in case of Silicon substrate. From Table 1 it is evident that there is a connection between Hurst exponent and crystallite size. In the early stages of thin film deposition, the Silicon substrate shows an anti-persistence effect. And, as the films grow the substrate effect goes on decreasing gradually till the growth displays persistence.

CONCLUSION

Higuchi algorithm is applied to the digitized AFM images for the estimation of fractal dimension of BaF_2 film deposited on Silicon, Aluminum and glass substrates, respectively. Our finding suggests that the growth of the film on glass substrate exhibits isotropic nature. Fractal dimension of BaF_2 films deposited on Silicon substrate is found to be higher as compared to that of the Aluminum as well as glass substrates which suggests that surface morphology of Silicon substrate is more irregular and complex. Our finding also suggests that Aluminum and glass substrate both show persistence effect while Silicon substrate shows anti-persistence effect. The small crystalline size could be responsible for the anti-persistence effect of silicon substrate. These finding suggest that the substrate strongly affects the surface morphology of the films.

CONSENT FOR PUBLICATION

Not applicable.

CONFLICT OF INTEREST

The authors declare no conflict of interest, financial or otherwise.

ACKNOWLEDGEMENT

Declared none.

REFERENCES

[1]   Mandelbrot B. Fractals and chaos: the Mandelbrot set and beyond. Springer Science & Business Media 2013.

[2]   Saitou M, Oshikawa W, Makabe A. "Characterization of electrodeposited nickel film surfaces using atomic force microscopy,"Journal of Physics and Chemistry of Solids, vol. 63, pp. 1685-1689, 2002/09/01/ 2002.

[3]   Snyder GJ, Hiskes R, DiCarolis S, Beasley MR, Geballe TH. Intrinsic electrical transport and magnetic properties of La_0.67Ca_0.33MnO_3 and La_0.67Sr$_{0.33}$MnO$_3$ MOCVD thin films and bulk material. Phys Rev B Condens Matter 1996; 53(21): 14434-44.
        [http://dx.doi.org/10.1103/PhysRevB.53.14434] [PMID: 9983241]

[4]   Aguilar ZP. Chapter 4 - Nanobiosensors. in: Aguilar ZP (Ed.) Nanomaterials for Medical Applications. Elsevier, 2013; pp. 127-79.

[5]   Lin M, Zhao Y, Wang S, *et al.* Recent advances in synthesis and surface modification of lanthanide-doped upconversion nanoparticles for biomedical applications," Biotechnology Advances, vol. 30, pp. 1551-1561, 2012/11/01/ 2012.

[6]   Venkatesan BM, Shah AB, Zuo J-M, Bashir R. DNA Sensing Using Nanocrystalline Surface-Enhanced Al$_2$O$_3$ Nanopore Sensors. Adv Funct Mater 2010; 20(8): 1266–75.

[7]   Monemdjou A, Ghodsi F E, Mazloom J. The effects of surface morphology on optical and electrical properties of nanostructured AZO thin films: Fractal and phase imaging analysis, Superlattices and Microstructures, vol. 74, pp. 19-33, 2014/10/01/ 2014.

[8]   Buzio R, Boragno C, Biscarini F, Buatier De Mongeot F, Valbusa U. The contact mechanics of fractal surfaces, Nat Mater, vol. 2, pp. 233-236, 04//print 2003.

[9]   Li J, Lü L, Lai MO, Ralph B. Image-based fractal description of microstructures. Springer Science & Business Media 2013.

[10]   Aguilar M, Oliva AI, Quintana P. The effect of electrical current (DC) on gold thin films. Surf Sci 1998; 409(3): 501-11.
        [http://dx.doi.org/10.1016/S0039-6028(98)00282-9]

[11]   Ţălu Ş, Bramowicz M, Kulesza S, *et al.* Microstructure and tribological properties of FeNPs@ aC: H films by micromorphology analysis and fractal geometry. Ind Eng Chem Res 2015; 54(33): 8212-8.
        [http://dx.doi.org/10.1021/acs.iecr.5b02449]

[12] Kavyashree Pandey RK, Yadav RP, *et al.*, Substrate effect on the evolution of surface morphology of BaF2thin films: A study based on fractal concepts, Applied Surface Science, vol. 466, pp. 780-786, 2019/02/01/ 2019.

[13] Thomas ME, Tropf WJ. Barium Fluoride (BaF2). in: Palik ED (Ed.) Handbook of Optical Constants of Solids, Burlington: Academic Press 1997: pp. 683-99.

[14] Sliney HE. Self-Lubricating Composites of Porous Nickel and Nickel-Chromium Alloy Impregnated with Barium Fluoride-Calcium Fluoride Eutectic. A S L E Transactions 1966; 9(4): 336-47.
[http://dx.doi.org/10.1080/05698196608972150]

[15] Gibbs WEK, Butterfield AW. Absorption of thin film materials at 106 μm. Applied Optics 1975; 14: pp. 3043-6.

[16] Ma D-a, Zhu R-y. Light attenuation length of barium fluoride crystals, Nuclear Instruments and Methods in Physics Research Section A: Accelerators, Spectrometers, Detectors and Associated Equipment, vol. 333, pp. 422-424, 1993/09/01/ 1993.

[17] Yadav RP, Kumar T, Mittal AK, Dwivedi S, Kanjilal D. Fractal characterization of the silicon surfaces produced by ion beam irradiation of varying fluences. Appl Surf Sci 2015; 347: 706-12.
[http://dx.doi.org/10.1016/j.apsusc.2015.04.150]

[18] Yadav RP, Kumar M, Mittal AK, Pandey AC. Fractal and multifractal characteristics of swift heavy ion induced self-affine nanostructured BaF_2 thin film surfaces. Chaos 2015; 25(8): 083115.
[http://dx.doi.org/10.1063/1.4928695] [PMID: 26328566]

[19] Lopes R, Betrouni N. Fractal and multifractal analysis: A review, Medical Image Analysis, vol. 13, pp. 634-649, 2009/08/01/ 2009.

[20] Mandelbrot BB. Self-Affine Fractals and Fractal Dimension. Physica Scripta 1985; 32: 257-60.
[http://dx.doi.org/10.1088/0031-8949/32/4/001]

[21] Higuchi T. Relationship between the fractal dimension and the power law index for a time series: A numerical investigation, Physica D: Nonlinear Phenomena 1990; 46(2): 254-64. 1990/11/01/ 1990.

[22] Pelliccione M, Lu T-M. Evolution of thin film morphology. Springer Series in Materials Science. vol. 108, ed: Springer, 2008.

[23] Shlesinger MF. Fractal Time in Condensed Matter. Ann Rev Phy Chem 1988; 39: 269-90.
[http://dx.doi.org/10.1146/annurev.pc.39.100188.001413]

[24] Liu SH. Fractals and Their Applications in Condensed Matter physics. In: Ehrenreich H, Turnbull D. Solid State Physics. Academic Press, 1986, 39: 207-73.
[http://dx.doi.org/10.1016/S0081-1947(08)60370-7]

[25] Yadav RP, Agarwal DC, Kumar M, Rajput P, Tomar DS, Pandey SN, *et al.*, Effect of angle of deposition on the Fractal properties of ZnO thin film surface, Applied Surface Science, vol. 416, pp. 51-58, 2017/09/15/ 2017.

[26] Ţălu Ş, Yadav RP, Mittal AK, Achour A, Luna C, Mardani M, *et al.*, Application of Mie theory and fractal models to determine the optical and surface roughness of Ag–Cu thin films, Optical and Quantum Electronics, vol. 49, p. 256, 2017/07/03 2017.

[27] Kesić S and Spasić SZ, Application of Higuchi's fractal dimension from basic to clinical neurophysiology: A review Computer Methods and Programs in Biomedicine, vol. 133, pp. 55-70, 2016/09/01/ 2016.

[28] Yadav RP, Pandey RK, Mittal AK, Kumar M, Pandey AC. Surface Roughness and Fractal Study of CaF_2 Thin Films. Materials Focus 2015; 4(6): 403-8.
[http://dx.doi.org/10.1166/mat.2015.1275]

[29] Alvarez-Ramirez J, Echeverria JC, and Rodriguez E, Performance of a high-dimensional R/S method for Hurst exponent estimation, Physica A: Statistical Mechanics and its Applications, vol. 387, pp.

6452-6462, 2008/11/15/ 2008.

[30] Yadav RP, Kumar M, Mittal AK, Dwivedi S, and Pandey AC, On the scaling law analysis of nanodimensional LiF thin film surfaces, Materials Letters, vol. 126, pp. 123-125, 2014/07/01/ 2014.

[31] Zhao Y, Wang G-C, Lu T-M. Characterization of Amorphous and Crystalline Rough Surface--Principles and Applications. Elsevier 2000.

Part 2: Nano Functional Materials and Their Application

A Detailed Study of Structural, Dielectric and Luminescence Properties of Sm^{3+} Doped $BiFeO_3$ Nanoceramics

Satish Kumar Mandal[1,2,*]**, Savita**[3,4,*]**, Pradip Kumar Priya**[5]**, Ram Pratap Yadav**[6]**, Hari Pratap Bhasker**[7]**, Raj Kumar Anand**[3] **and Amreesh Chandra**[1]

[1] *Department of Physics, Indian Institute of Technology Kharagpur, Kharagpur-721 302, India*

[2] *Surface Physics and Material Science Division, Saha Institute of Nuclear Physics, Kolkata-700064, India*

[3] *Department of Physics, University of Allahabad, Prayagraj-211 001, India*

[4] *Government Girls Polytechnic Ayodhya, Uttar Pradesh-224001, India*

[5] *Department of Physics, Ewing Christian College, University of Allahabad, Prayagraj-211 003, India*

[6] *Department of Physics, Deen Dayal Upadhyay Govt. P.G. College, Saidabad, Prayagraj-221 508, India*

[7] *Department of Physics, Chaudhary Mahadeo Prasad Degree College, University of Allahabad, Prayagraj-211 002, India*

Abstract: Observation of at least two coexisting switchable ferroic states *viz.*, ferromagnetic, ferroelectric, and/or ferroelastic at room temperature with promising coupling among order parameters, has made $BiFeO_3$ a highly explored material in the field of multiferroics and/or magnetoelectric multiferroics, which creates the possibility for its application in various technological devices such as spintronics, spin-valve, DRAM, actuators, sensors, solar-cells photovoltaic, *etc.* Intrinsically, its low coupling coefficients, difficulty to prepare in pure phase in bulk, high leakage current, *etc.* have restricted $BiFeO_3$ from technological reliability. However, the effect of doping with iso- and alio-valent ions, nanostructure, thin-film-form and nanoparticles, *etc.*, has been carried out to improve its physical properties by several research groups over the decades. In this chapter, the structural, luminescence, and dielectric properties of samarium (Sm^{3+}) doped $BiFeO_3$ nanoceramics synthesized using a modified gel-combustion route are discussed in detail. The effect of Sm^{3+} doping in $BiFeO_3$ is explored using the X-ray diffraction (XRD) technique. The XRD studies exhibit a possible structural phase transition above Sm^{3+} doping of 15% from rhombohedral (R3c) space group to the orthorhombic (Pbnm) space group. The dielectric study shows interesting behavior accompanied by structural transition. Our study suggests that Sm^{3+}

---

* **Corresponding authors Satish Kumar Mandal and Savita:** Department of Physics, Indian Institute of Technology Kharagpur, Kharagpur-721 302, India; and Government Girls Polytechnic Ayodhya, Uttar Pradesh-224001, India; E-mails: satishbhu17@gmail.com and 01savita@gmail.com

Arti Srivastava, Mridula Tripathi, Kalpana Awasthi and Subhash Banerjee (Eds.)

doping plays an important role in governing the structural, luminescence, and dielectric properties of $BiFeO_3$ samples.

Keywords: Bismuth Ferrite (BFO), Multiferroic, Dielectric Properties.

INTRODUCTION

In the last few decades, the world has been witnessing a quantum leap in the digitalization of data in day to day routine and storage devices [1 - 3]. As a result, rapid advancement has also been seen in the field of memory storage electronics devices and their allied area of technologies. The current global demand is a storage device which is fast, and robust with low power consumption. In general, the memory devices fundamentally use ferroelectric and ferromagnetic materials for data writing and operations [1, 3, 4]. The ferroelectric (FE) or ferromagnetic (FM) materials have been extensively used in memory devices due to their switchable polarization (magnetization) states [1, 4, 5]. In commercial magnetic memory device (MRAM), the data is stored (or write operation) by switching the magnetic states from $-M$ to $+M$ or vice versa while, the information can be read by changing the magneto resistance. Currently, the memory devices fabricated using ferromagnetic materials with high coercivity consume a lot of energy for switching from one state to the other. In ferroelectric random access memory (FeRAM) devices, the switching from one state to another state can be done by simply changing the state of polarization. FeRAM devices show a faster writing speed and consume a comparatively smaller amount of energy [1, 4, 5]. But, these devices show some limitations like slow reading operation [3].

Therefore, it is necessary to find a cost-effective suitable material in which the above properties simultaneously coexist and reduce the processing steps with better. Interestingly, the multiferroic materials are the materials in which two and/or more switchable ferroic order parameters such as Ferroelectricity, ferromagnetism and /or Ferroelastic exist simultaneously [6, 7]. Sometimes, multiferroics showing magnetoelectric properties through direct and /or indirect coupling coefficients prove to be quite useful for storage applications [8]. In such magneto-electric (ME) multiferroic materials, the spontaneous magnetization (M_s) and /or spontaneous polarization (P_s) can be switched by controlling external electric and /or magnetic fields [3]. In both cases, the spontaneous deformation can also be switched /reoriented by applying stress (σ) [9, 10]. These materials show many technological prospects in the field of switching devices, novel memory media, transducers, new functional sensors *etc.*, [2, 11]. Hitherto, functional devices made of multiferroic materials are yet to be realized [12, 13]. There are only few materials like $BiFeO_3$ [14], $TbMnO_3$ [15], $BiMnO_3$ [16] and

$YMnO_3$ [17] *etc.*, that exhibit multiferroic properties and, therefore the current demand is to synthesize and explore the new materials for their application in multifunctional devices. Among these materials, $BiFeO_3$ (BFO) is the only material showing prominent multiferroism at room temperature [2, 6]. BFO has a distorted rhombohedral crystal structure with R3c space group symmetry which permits anti-phase octahedral distortion and ionic displacement from the centro-symmetric position. The BFO shows ferroelectric property at room temperature with very high Curie temperature [$(T_C) \sim 1103K$] and G-type anti-ferromagnetism up to Neel temperature ($(T_N) \sim 640K$) [18, 19]. Moreover, in BFO, the neighboring magnetic spins are oriented antiparallel to each other. In addition, it makes a spiral spin structure with a large cycloidal period of ~620 Å precise throughout the crystal [20].

In order to obtain intriguing properties of the BFO, several direct and indirect methods can be employed. The multiferroic properties can be improved by substitution of rare earth (lanthanide series) elements at A-site (*i.e.* Bi site) and transition metals at B-site (*i.e.*, Fe site), respectively [21, 22]. BFO has been explored for their application in photo-catalytic activity for the degradation of organic pollutants [23]. There are reports that Sm and Mn doping enhances the photocatalytic activity of $BiFeO_3$ nanoparticles [24]. Lou *et al.* reported that $SrTiO_3$-coated BFO core-shell nanostructures can be employed as a water splitter under visible-light illumination [23, 25].

Here, we have discussed the synthesis and characterization of BFO and Sm doped BFO nano ceramics. All the samples are prepared using the modified gel-combustion route. The effects of Sm doping in BFO on structural, luminescence and dielectric properties are systematically investigated. The structural analysis of the samples suggests that the suitable doping of Sm in BFO plays an important role in governing structural, luminescence and dielectric properties.

EXPERIMENTAL PROCEDURE

Ceramics samples of $Bi_{1-x}Sm_xFeO_3$ ($0.0 \leq x \leq 0.40$) were synthesized using a modified gel-combustion route. The details can be found elsewhere [26]. To start, a stoichiometric ratio of $Bi(NO_3)_3 \cdot 5H_2O$ (99.0% purity, MERCK, India), $Sm(NO_3)_3$ (96%, purity, MERCK, India) and $Fe(NO_3)_3 \cdot 5H_2O$ (99% purity, MERCK, India) was dissolved in distilled water. Citric acid (99% purity, MERCK, India) was used for complexation as well as a fuel agent. The molar ratio of bismuth to citric acid was maintained at 1:1.5. Due to the different pH of individual solutions, it was difficult to obtain a clear solution on mixing. Therefore, dilute HNO_3 was added drop wise for precipitate dilution until the murky solution transformed into a clear solution. The obtained solution was then continuously stirred for 2 hours at constant temperature ~70°C till the solution became a transparent yellow color. In order to prevent the agglomeration and particle sedimentation, 25 ml of polyethylene glycol 400 (PEG-400) was added while the solution was being continuously stirred. PEG plays the role of a colloidal suspension medium. Finally, after stirring for 6 to 8 hours, a highly viscous brownish coloured gel could be obtained. Further, to remove the solvents, the gel was heated at 250°C and this process yields a brownish porous ash-like material. The structural properties of the as-prepared samples (with x = 0.05; from now denoted as BSmFO 05) calcined at 550°C temperatures for 6 hours was carried out using X-ray diffraction (XRD). The calcined powders were pelletized in the form of 8 mm diameter circular discs using a hydraulic press (KBr press, KIMAYA, India) under a pressure of 4 tons. 2% PVA was used as a binder. The pellets were sintered at 750°C for 6 hours to obtain dense ceramic samples [26]. For x-ray diffraction (XRD) studies, the high-density sintered pellets were crushed using a mortar pestle. The crushed powders were annealed for 10 h at 500°C to remove the residual strain. XRD data was collected using an X'Pert Pro diffractometer in the 2θ range 20°–110°. Photoluminescence spectra for all samples were recorded using continuous wave He-Cd laser with excitation wavelength 325 nm.

RESULT AND DISCUSSION

Structural Analysis

Fig. (1) shows the X-ray diffraction (XRD) patterns for pure and Sm doped BFO samples, respectively. Both the samples exhibit typically rhombohedral distorted perovskite structure with R3c space group. The structural phase changes from rhombohedral distorted perovskite structure with R3c space group to orthorhombic structure with Pbnm space group at Sm doping concentration of 15% and beyond [Rietveld refinement of the XRD data is not shown here]. The impurity phase corresponding to $Bi_2Fe_4O_9$ is also present (labeled as*). The characteristic rhombohedral peaks have been shifted to higher diffraction angles suggesting possible structural distortions upon Sm^{3+} doping at highly polarizable aliovalent site of Bi^{3+} in pure BFO which might have occurred due to the difference in ionic radii of these two cations which are 1.17Å and 1.107Å for Bi^{3+} and Sm^{3+}, respectively.

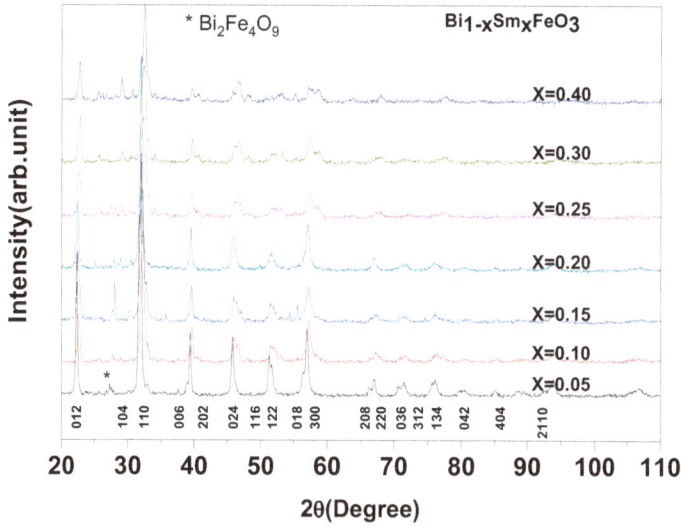

Fig. (1). Room temperature XRD pattern of $Bi_{(1-x)}Sm_{(x)}FeO_3$ ($0.0 \leq x \, 0.4$) nanoceramics.

Microstructural Analysis

Fig. (**2a-b**) shows shows the top view SEM micrograph for $Bi_{0.95}Sm_{0.05}FeO_3$ and (c-d) for $Bi_{0.85}Sm_{0.15}FeO_3$ samples, respectively. The size of the particles shows a large variation in size, and all the samples show agglomeration effect.

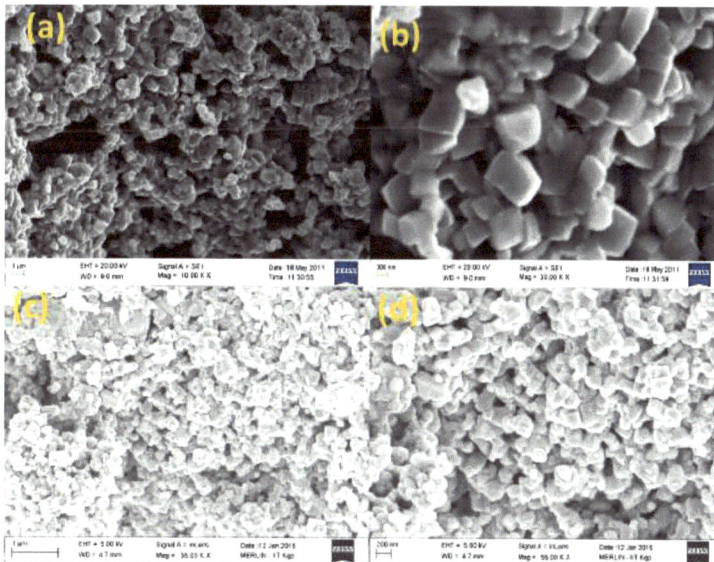

Fig. (2). Top view SEM micrograph for (a-b) $Bi_{0.95}Sm_{0.05}FeO_3$ and (c-d) $Bi_{0.85}Sm_{0.15}FeO_3$.

Luminescence Properties

Fig. (**3**) shows the photoluminescence (PL) spectra for as prepared $BiFeO_3$ sample. The PL spectra of the pure BFO consist of several peaks. The band edge peaks is not visible and it is masked with some unknown defects related emission. A strong blue emission peak appears at the wavelength of ~402 nm. This emission could be due to the self-activated centers in the synthesized nanoparticles. The peak appears at ~649.2 nm may be due to onsite Fe^{3+} crystal field.

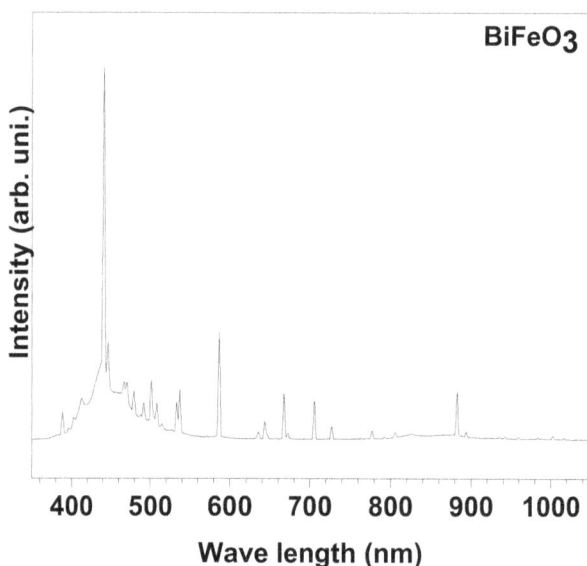

Fig. (3). Photoluminescence spectra for $BiFeO_3$ samples recorded using He-Cd laser with excitation wavelength of 325 nm.

Dielectric Properties

Fig. (**4**) shows the variation in dielectric constant (ε_r) as a function of temperature for BSmFO 05 sample at different frequencies ranging from 100Hz to 1MHz. As the temperature increases, the value of dielectric constant increases and goes through a peak which appears at 283°C. The sample shows an increasing trend in dielectric permittivity over a wide temperature range suggesting normal ferroelectric and/or ferromagnetic behaviour. This observed increment in dielectric permittivity upon samarium (Sm^{3+}) doping compared to pure BFO sample might be due to well-developed grains, increased density and less oxygen deficiency present in the sample. The characteristic frequency independent of peak maxima (T_m) and the vicinity of magnetic Neel temperature ($T_N = 380°C$)

has been observed showing dielectric anomaly coupled with magnetic transition. A significant deviation in magnetic anomaly peak compared to $T_N = 380°C$ of pure $BiFeO_3$ sample has been observed in our sample compared to other reported values. Apart from this magnetically coupled dielectric anomaly, two more dielectric anomalies at $T = 283°C$ and $T = 30°C$ (close to the room temperature) have also been seen as depicted in the inset of Fig. (**1**).

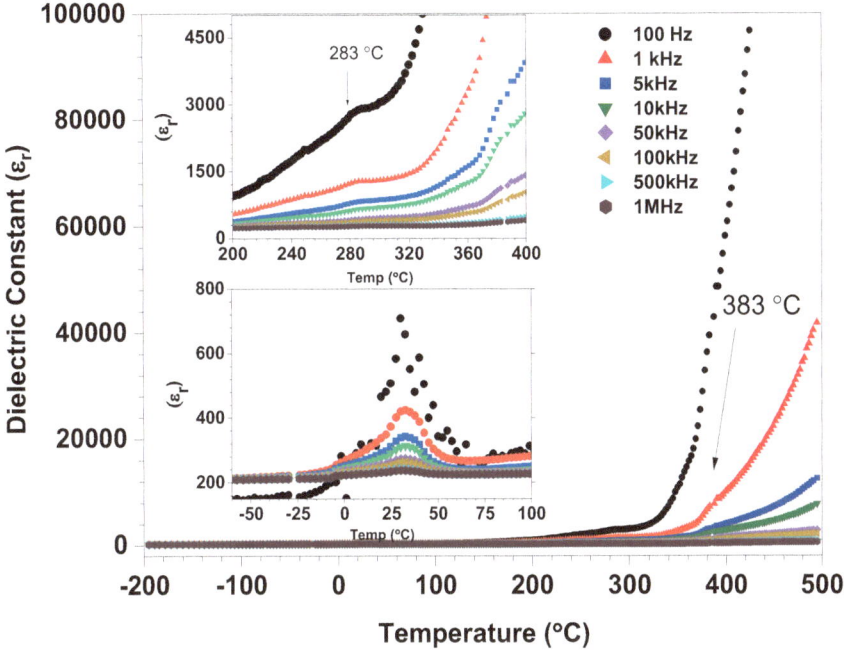

Fig. (4). The variation in dielectric constant for 5% Sm doped $BiFeO_3$ as a function of temperature recorded at different frequency. Inset to figure shows the zoom view of peak around room temperature (T=30°C) and another at T=283°C for clarity.

Fig. (**5**) shows the variation in tangent loss (tanδ) for 5% Sm doped $BiFeO_3$ as a function of temperature recorded at different frequency. The tangent loss increases as temperature increases along with peaks at 30°C and 300°C. Both the peaks are frequency independent. The value of tanδ is also decreases as frequency increases. However, the magnitude of dielectric loss tanδ shoots up beyond 400°C which might be due to the dominant space charge polarization effect associated with thermally activated oxygen vacancies in the sample at a high temperature.

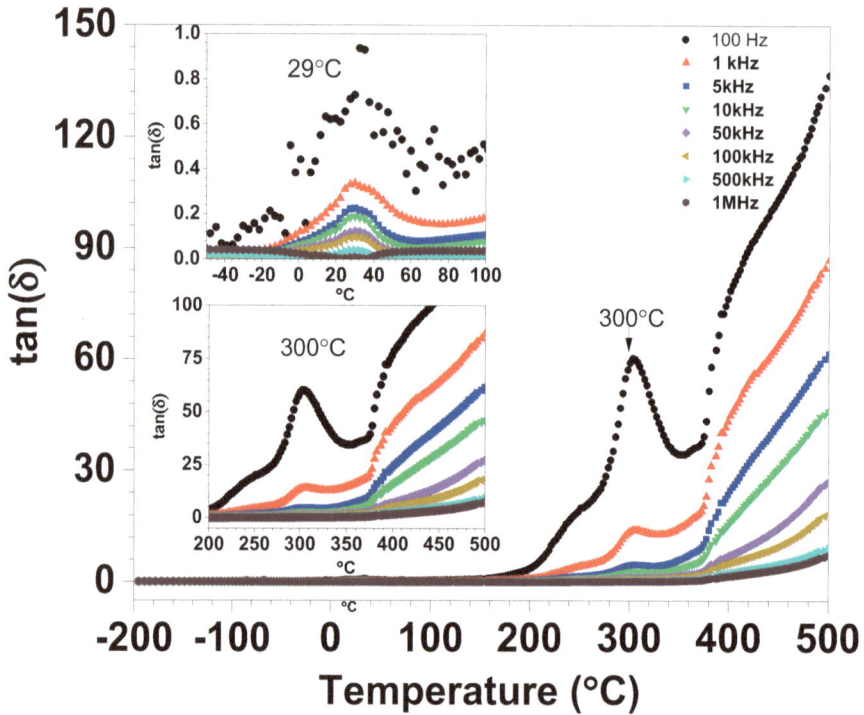

Fig. (5). The variation of tangent loss [tan(δ)] for 5% Sm doped BiFeO₃ as a function of temperature recorded at different frequency.

CONCLUSION

The pure and Sm-doped $BiFeO_3$ (BFO) samples are synthesized by a modified sol-gel route. The effect of doping on the structural, luminescence and dielectric properties of these samples is systematically investigated. The dielectric study shows interesting behavior. The structural analysis of the samples suggests that the suitable doping of Sm in BFO plays an important role in governing structural, luminescence and dielectric properties.

CONSENT FOR PUBLICATION

Not applicable.

CONFLICT OF INTEREST

The authors declare no conflict of interest, financial or otherwise.

ACKNOWLEDGEMENTS

The first author thankfully acknowledges DAE-BRNS, BARC (Mumbai) for funding under the BRNS-Young Scientist Research Award Scheme.

REFERENCES

[1]   Alexe M, Gruverman A, Harnagea C, *et al.* Switching properties of self-assembled ferroelectric memory cells,"Applied Physics Letters vol. 75, pp. 1158-1160, 1999/08/23 1999.

[2]   Ramesh R, Spaldin NA. Multiferroics: progress and prospects in thin films," Nature Materials vol. 6, pp. 21-29, 2007/01/01 2007.

[3]   Wang J. Multiferroic Materials: Properties, Techniques, and Applications. CRC Press 2016.
       [http://dx.doi.org/10.1201/9781315372532]

[4]   Scott, J. Multiferroic memories. Nature Mater 2007; 6: 256–57.
       [http://dx.doi.org/10.1038/nmat1868]

[5]   Kohlstedt H, Mustafa Y, Gerber A, *et al.* Current status and challenges of ferroelectric memory devices Microelectronic Engineering vol. 80, pp. 296-304, 2005/06/17/ 2005.

[6]   Fiebig M, Lottermoser T, Meier D, Trassin M. The evolution of multiferroics Nature Reviews Materials vol. 1, p. 16046, 2016/07/05 2016.

[7]   Hill N A. Why Are There so Few Magnetic Ferroelectrics? The Journal of Physical Chemistry B vol. 104, pp. 6694-6709, 2000/07/01 2000.

[8]   Nan C-W, Bichurin M I, Dong S, Viehland D, Srinivasan G. Multiferroic magnetoelectric composites: Historical perspective, status, and future directions Journal of Applied Physics vol. 103, p. 031101, 2008/02/01 2008.

[9]   Singh G, Bhasker H P, Yadav R P, *et al.* Dielectric, magnetic and magneto-dielectric properties of (La, Co) co-doped $BiFeO_3$ Physica Scripta vol. 94, p. 125805, 2019/09/20 2019.

[10]   Tirupathi P, Mandal S K, Chandra A. Effect of oxygen annealing on the multiferroic properties of Ca^{2-} doped $BiFeO_3$ nanoceramics Journal of Applied Physics vol. 116, p. 244105, 2014/12/28 2014.

[11]   Němec P, Fiebig M, Kampfrath T, Kimel AV. Antiferromagnetic opto-spintronics. Nature Phys 2018; 14; 229–41.
       [http://dx.doi.org/10.1038/s41567-018-0051-x]

[12]   Johnson RD, Radaelli PG. Diffraction Studies of Multiferroics Annual Review of Materials Research 2014; 44(1): 269-98.
       [http://dx.doi.org/10.1146/annurev-matsci-070813-113524]

[13]   Padmanabhan H, Munro JM, Dabo I, Gopalan V. Antisymmetry: Fundamentals and Applications Annu Rev Mat Res 2020; 50: 255-81.

[14]   Catalan G, Scott JF. Physics and Applications of Bismuth Ferrite Adv Mater 2009; 21(24): 2463-85.
       [http://dx.doi.org/10.1002/adma.200802849]

[15]   Kimura T, Goto T, Shintani H, Ishizaka K, Arima T, Tokura Y. Magnetic control of ferroelectric polarization. Nature 2003; 426(6962): 55-8.

[16]   Béa H, Gajek M, Bibes M, Barthélémy A. Spintronics with multiferroics. J Phys Condens Matter 2008; 20(43): 434221.
       [http://dx.doi.org/10.1088/0953-8984/20/43/434221]

[17]   Van Aken BB, Palstra TTM, Filippetti A, Spaldin NA. The origin of ferroelectricity in magnetoelectric YMnO3. Nat Mater 2004; 3(3): 164-70.

[18]   Kubel F, Schmid H. Structure of a Ferroelectric and Ferroelastic Monodomain Crystal of the

Perovskite BiFeO₃. Acta Crystallogr Sect B: Struct Sci 1990; 46(6): 698-702.
[http://dx.doi.org/10.1107/S0108768190006887]

[19] Fischer P, Polomska M, Sosnowska I, Szymanski M. Temperature dependence of the crystal and magnetic structures of BiFeO ₃. J Phys C Solid State Phys 1980; 13(10): 1931-40.
[http://dx.doi.org/10.1088/0022-3719/13/10/012]

[20] Sosnowska I, Neumaier TP, Steichele E. Spiral magnetic ordering in bismuth ferrite. J Phys C Solid State Phys 1982; 15(23): 4835-46.
[http://dx.doi.org/10.1088/0022-3719/15/23/020]

[21] Yang CH, Kan D, Takeuchi I, Nagarajan V, Seidel J. Doping BiFeO₃: approaches and enhanced functionality. Phys Chem Chem Phys 2012; 14(46): 15953-62.
[http://dx.doi.org/10.1039/c2cp43082g] [PMID: 23108014]

[22] Wu J, Wang J. Ferroelectric and impedance behavior of La□and Ti□codoped BiFeO₃ thin films. J Am Ceram Soc 2010; 93(9): 2795-803.
[http://dx.doi.org/10.1111/j.1551-2916.2010.03816.x]

[23] Gao F, Chen XY, Yin KB, *et al.* Visible-Light Photocatalytic Properties of Weak Magnetic BiFeO₃ Nanoparticles. Adv Mater 2007; 19(19): 2889-92..

[24] Irfan S, Shen Y, Rizwan S, Wang H-C, Khan S B, Nan C-W. Band-Gap Engineering and Enhanced Photocatalytic Activity of Sm and Mn Doped BiFeO₃ Nanoparticles. J Amer Ceram Society 2017; 100(1): 31-40.

[25] Luo J, Maggard PA. Hydrothermal Synthesis and Photocatalytic Activities of SrTiO₃-Coated Fe₂O₃ and BiFeO₃. Adv Mater 2006; 18(4): 514-7..

[26] Mandal SK, Rakshit T, Ray SK, Mishra SK, Krishna PSR, Chandra A. Nanostructures of Sr2+ doped BiFeO₃ multifunctional ceramics with tunable photoluminescence and magnetic properties. J Phys Condens Matter 2013; 25(5): 055303.

CHAPTER 7

Application of Nanotechnology in Wastewater Cleaning Process

Monika Singh[1,*], Deepanjali Pandey[1], Dharamveer Singh[2], Shalini Verma[3] and **Vijay Krishna[3]**

[1] *Department of Chemistry, CMP Degree College, University of Allahabad, Uttar Pradesh 211002, India*

[2] *School of Sciences, Rajshree Tondon Open University, Prayagraj, India*

[3] *Department of Chemistry, University of Allahabad, Prayagraj, India*

Abstract: With the advancement in research, new techniques are growing very fast these days. The environmental contamination by many hazardous elements is seen in today's world. The radioactive materials and their byproducts or the leakage of nuclear reactors is a potential serious health threat. The ground water and drinking water get contaminated and it is a big challenge to remove these radioactive ions from the environment. The radioactive ions leach into groundwater and contaminate drinking water supplies for large population areas. The key issue in developing technologies for the removal of radioactive ions from the environment mainly from wastewater and their subsequent safe disposal is to devise materials which are able to absorb radioactive ions irreversibly, selectively, efficiently, and in large quantities from contaminated water. Hence, nanotechnology proved to be a great success in this area. Nanotechnology is the science and technology working at the molecular level *i.e.* in nanometre and embraces many different fields and specialties, including engineering, chemistry, electronics, medicine, pharmaceuticals, agriculture and waste management. The present chapter deals with the development of nano-technology for the removal and safe disposal of radioactive ions from the environment using nanomaterials.

Keywords: Nanotechnology, Nanomaterials, Radio Wastes, Radionuclides.

INTRODUCTION

Nanotechnology is one of the fastest and widest fields of science in this decade. New inventions using nanomaterials as a tool prove to be a boon to society, *i.e.* research and technology development at the atomic, molecular or macromolecular levels, approximately 1-100 nanometers in length. All nanomaterials of this size

---

* Corresponding author Monika Singh: Department of Chemistry, CMP Degree College, University of Allahabad, Uttar Pradesh 211002, India; E-mails: monikasingh09_inn@rediffmail.com, msrajharsh@gmail.com

Arti Srivastava, Mridula Tripathi, Kalpana Awasthi and Subhash Banerjee (Eds.)

can control and manipulate the atomic scale since the properties are different from the bulk materials. We know many radioactive materials that help our society, and improve the quality of life, ranging from power generation to industrial uses, medical, diagnostic, therapeutic, and research purposes.

In medicine, a number of unsealed sources are used for diagnosis while sometimes they need to be diluted before use. Also, a number of biomedical researchers use radionuclides which may increase the volume of wastes requiring further management. Radioactive waste is mainly a byproduct of all such nuclear power generation and is strictly regulated as a highly hazardous material to human beings as well as the environment. Radioactive materials create ions, which initiate free radical formation. These free radicals can damage protein membrane and nucleic acids by "oxidizing" them generally known as "OXIDATIVE STRESS". Many forms of cancer are thought to be the result of reactions between free radicals and DNA, causing mutations that can adversely affect the cell cycle and potentially lead to malignancy. Leaching of such radiowastages and byproducts into groundwater can contaminate drinking water supplies is a serious health threat.

Today, newer generation nanomaterials as functionalized nanomaterials proved to be very effective for radioactive wastewater decontamination, such as carbon-based nanomaterials, metal nanoparticles, nano-sized metal oxides, metal sulfides, nano-sized natural materials, layered double hydroxides, hydroxyapatite nanoparticles, metal-organic frameworks, cellulose nanomaterials, and biogenic nanocomposites. Functionalized materials can effectively remove specific ions of interest, in chemical separation methods. Recent progress in the synthesis of nano-structured materials helps to remediate radioactive contamination at the source, by removing radioactive ions from the different source of contamination. There are several methods that can use nanotechnology as a reactive media for separation and filtration, bioremediation and disinfection [1 - 3]. Some of them are discussed below:

INORGANIC CATION EXCHANGE MATERIALS- MATERIAL FOR WATER CLEAN UP

"Natural inorganic cation-exchanger materials, such as clay and zeolite, have been extensively studied and used in the removal of radioactive ions from water *via* ion exchange and are subsequently disposed of in a safe way", Dr. Huai Yong Zhu [4] explains to "Nanowerk". According to Dr. Zhu, the Queensland University of Technology in Brisbane, Australia, "Synthetic inorganic cation exchange materials - such as synthetic micas, g-zirconium phosphate, niobate molecular sieves, and titanate - have been found to be far superior to natural materials in terms of selectivity for the removal of radioactive cations from water. Generally, ion exchange materials exhibiting a layered structure are less stable than those with 3D crystal structures and the collapse of the layers can take place under moderate conditions" says Zhu. "Then again, it has also been found that nanoparticles of inorganic solids readily react with other species or are quickly converted to other crystal phases under moderate conditions, and thus are substantially less stable than the corresponding bulk material". The uptake of large, radioactive cations eventually triggers the trapping of the cations - by itself represents a desirable property for any material to be used in the decontamination of water having radioactive cations (Fig. **1**).

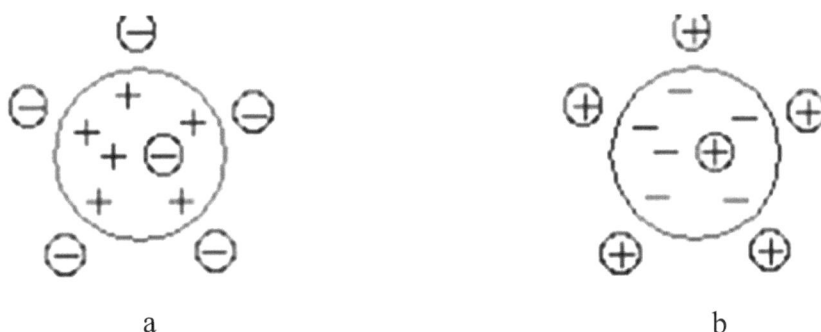

a   b

Fig. (1). a) Anion exchanger - attracts negatively charged particles and collects them and **b)** Cation exchanger - attracts positively charged particles and collects them.

Zeolites have properties such as high cation exchange capacities, high specific surface areas (due to porous crystalline structure) and high hydrothermal stability [5]. Zeolites have the ability to remediate water containing cationic species, such as ammonium and heavy metal ions also. Radioactive species drained out from the nuclear plants as wastewater and polluted groundwater such as ^{137}Cs and ^{90}Sr can be treated using these zeolites. For synthesizing zeolites, conventionally on the scale of 1,000 to 10,000 nm or 1 to 10 μm but since the pores size range between 0.4 to 1 nm and hence, zeolites are considered nanomaterials. Recently, nanocrystalline zeolites are being synthesized for discrete, uniform crystals with dimensions of less than 100 nm [6].

IRON NANOMATERIALS

Iron nanomaterials are prominently used for cleaning radiowaste-water. 'Pump and Treat' system [7] and Permeable Reactive Barrier (PRB) are the two techniques used. 'Pump and Treat' is the most widely used method since 1998 for wastewater treatment where the underground water is pumped out, treated and inserted back into the ground.

Another method to remediate water is by Permeable Reactive Barrier (PRB) system used to clean up pollutants such as chlorinated hydrocarbons, polychlorinated biphenyls (PCBs), pesticides, chromate compounds, and aromatic nitro compounds. Here, the subsurface water is cleaned without bringing out the water, proving to be time and energy-saving. In the 1990s, many zero-valent ions (ZVI) of metals such as iron [8] were actively used in PBR TECHNIQUE. These nano-size irons (nZVI) with their reductive potentials acted multifunctional to control and reduce the toxic contaminants in water. Hence, large quantities of water can be treated more effectively in less time with ZVI as a filter material of PBR. nZVI has a reductive potential, higher reactivity towards a broad range of contaminants including halogenated compounds, nitrate, phosphate, polycyclic aromatic hydrocarbons, and heavy metals, and a higher mobility compared to its microscale counterpart. Consequently, nZVI is regarded as a promising remediation strategy suitable to a broad range of applications and environments. Still there are drawbacks related to nZVI, like 1) magnetic attraction between nano iron particles causes the rapid aggregation of particles, 2) nZVI are prone to react with dissolved oxygen and oxygen-rich compounds and 3) nZVI are toxic towards microbial species including lipids, proteins, and nucleic acids, damaging intracellular structures and eventually leading to cell death. Under aerated conditions, Fe^{2+} oxidizes more rapidly than under anaerobic conditions, the contribution of Fe^{2+} to nZVI toxicity is higher under anaerobic conditions than under aerobic conditions [9]. Modification of nZVI, *e.g.* as a result of nZVI coating with poly-asparaginate, reduces its toxicity by limiting the direct contact of nanoparticles with cells. Nowadays, nZVI particles are immobilized in or on suitable solid supports (*e.g.* other metal doping, coating the surface, or deposition on the support) to expand the effective pH range of the Fenton reaction and help overcome the drawbacks of nZVI. There are other metals such as zinc and tin having the property to decontaminate waste water [10]. Two metal alloys such as iron and iron-nickel-copper have been employed to degrade trichloro ethene and trichloro ethane [11]. The commonly used metals are palladium, silver, platinum, cobalt, copper and gold, while aluminum is used as inert. Another example is iron-platinum particles, which possess similar capabilities in degrading chlorinated benzenes [12] present in waste-water.

FERRITIN

Ferritin is an intracellular protein found in bacteria, algae, higher plants, and animals. It helps in storage and release of iron as required by the body (Fig. **2**).

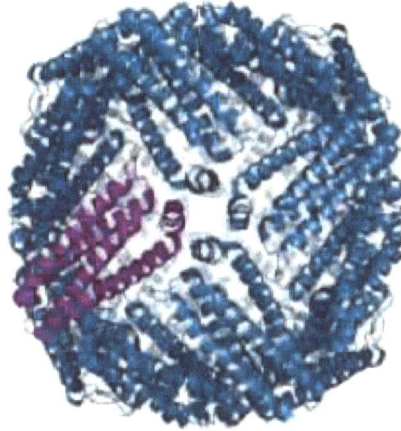

Fig. (2). Structure of a ferritin complex.

Adsorption is recognized as an efficient and economical method employed in water treatment [13]. For this, ferritin showed excellence in rapid phosphate and arsenate removal. These ferritins are thermally stable and give products at 10-12 m range. Radioisotopes ^{32}P and [76]As present as oxo-anions can be frequently removed using ferritin. It is an iron storage protein that is universally found in prokaryotes and eukaryotes. The ferritin is stable and active at 100°C for 10 hr and resists sterilization at 121°C for 30 min without the loss of enzymatic activity [14].

Ferric iron nanoparticles encapsulated in a ferritin nanocage act as a sorbent for orthophosphate, forming an iron-oxyhydroxide-phosphate nanoparticle. The mineral core can be regenerated by the release of phosphate after iron reduction [15]. In addition to PO_4^{3-} this protein has a high capacity for removal of other oxoanions such as arsenate AsO_4^{3-} or vanadate VO_4^{3-} [16] with the removal capacity of approximately 11 mg PO_4^{3-} per gram ferritin and 7 mg AsO_4^{3-}/g ferritin. As compared to Pump and Treat system, installation costs for PRBs are relatively high and the effective "time to replacement" is often uncertain [17].

COLLOIDAL NANOPARTICLES

Nanowaste management can be done in two ways: 1) once through or dry storage process and 2) closed cycle or aqueous dissolution process.

The benefit of using cementitious materials in the design of most of the Geological disposal facilities is known these days. At the pH value range 10-13, cement leachates even where the U(VI) concentration is very low around 10^{-9}M [18, 19]. Cement related phases just like iron oxides and silicates which will be ubiquitous in the GDF and the surface charges of colloidal nanoparticles could facilitate the transport of U(VI) into the geosphere [20, 21]. There are studies revealing cement and sediment-derived colloidal particles with complexed radionuclides [20] and intrinsic radionuclide colloids, *e.g.*, ThO_2 [21] and $Pu(OH)_4$ [22] as well as uranates, that act as a transport vector. Radioactive spills contaminating large areas are difficult to store and require increased safety measures. The radioactive isotope of Phosphorus *i.e.*[32]P is generally used to trace the pathway of biochemical reactions. For radioactive waste water treatment, magnetite is used for THE adsorption of phosphates and removal of phosphorus. Vinod *et al.* [23] synthesized and characterized a Karaya (a composite of gum and magnetite) to remove radioactive phosphates from ^{32}P-labeled biomolecules. The biodegradable hydrocolloid-based materials are efficient in radionuclide remedies.

CARBON NANOTUBES

Chemical impregnation of activated carbon is a good remedy for radiowaste-water. But carbon nanotubes (CNT), buckypapers, and other CNT composite materials showed much better response in the clean-up process. Multiwalled CNTs can adsorb some lanthanides and actinides like Eu (III) [24], Am(III) [25], and Th (IV) [26, 27], and radio-wastages present in water (Fig. **3**).

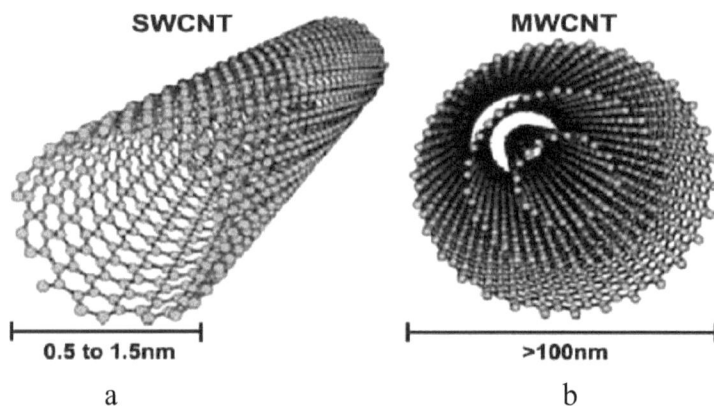

SWCNT   **MWCNT**

0.5 to 1.5nm   >100nm

a   b

Fig. (3). Schematic representation of **a)** Single Walled Carbon Nanotube (SWCNT) and **b)** Multi Walled Carbon Nanotube (MWCNT).
https://www.researchgate.net/publication/319966218_Carbon_nanotube_buckypaper_reinforced_polymer_co mposites_A_review/Fig.s?lo=1

CNTs are poorly soluble in water but functionalized CNT through oxidation can drastically increase the solubility and binding affinity of CNT to those of uranyl molecules [28, 29]. 18-crown- [6] ether and CNT interact through non-covalent type interaction and increase the solubility. Similar findings were investigated by Schierz and Zanker [30] used functionalized SW-CNT at semi-neutral pH for uranyl adsorption. Recently, researchers found a new set of macro-cyclic compounds named cucurbituril (CB), which have enhanced affinities to cations [31] and more cation binding affinities than crown-ethers and cyclodextrins [32]. And later adjusting the pH, uranyl bound CB molecule can be unbounded from the CNT molecule [33] (Fig. **4**).

Fig. (4). Structure of nano-fibres.

TITANATE NANOFIBERS

Nowadays, titanate nano-fibres are one of the intelligent absorbents for the remediation of radioactive ions from the contaminated water. These nano-fibres successfully entrap a large number of bivalent radioactive cations (approximately 80%) permanently and thus the structure of these nano-fibres gets deformed and further can be discarded. Actually, titanate materials are unique carriers of radioactive ions. The reason is their stability against radiation, chemicals, mechanical stress and thermal resistivity. These nano-fibers exhibit many advantages over clays and zeolites water purification techniques. Being fibrous in nature, they do not aggregate in water-samples and can be dispersed in the solution easily. Also, they can be easily removed from the water after use by any technique viz. filtration, sedimentation, centrifugation, *etc.* For these reasons, titanate nanofibers can be used on an industrial scale for the treatment of radioactive water wastes.

FERRIC HEXACYANOFERRATE $Fe_4[Fe(CN)_6]_3 \cdot XH_2O$

Ferric(III) hexacyanoferrate(II), a pigment of dark blue color and was one of the first synthetic dyes, also known as Prussian blue (Fig. **5**).

Fig. (5). Ferric(III) hexacyanoferrate(II) commonly known as Prussian blue.

Radioactive Cesium (^{137}Cs) is water soluble and behaves similar to potassium and sodium in the biological behavioural profile. This radio-waste can be removed by the use of Prussian blue. It is a low-cost adsorbent, with high selectivity for cesium, including a high stability, high conductivity, biocompatibility, size controllability, and easy surface functionalization for decomposition [34]. Cesium affect human body and can lead to carcinoma of the liver, kidney, bladder, impaired renal functions, cardiovascular disease, and gastrointestinal distress [35 - 37]. It is also an FDA-approved drug used in the treatment of radioactive exposure [38 - 40].

NANOFILTRATION

Nanofiltration is a filtration process where pressure acts as a driving force through nano- membranes and targeting multivalent ions based on their molecular weight, pesticides and heavy metals can be filtered out. It is a very economical process, and one of the newest and most leading-edge technologies of water treatment [41, 42] (Fig. **6**).

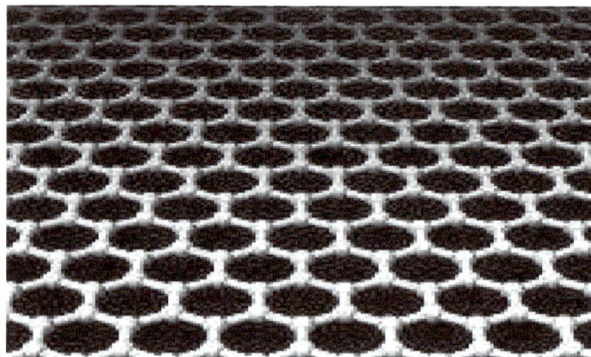

Fig. (6). Graphene coated nanofilter membrane.

Nanofiltration method is used in softening water and retaining calcium and magnesium ions. While passing smaller hydrated monovalent ions, filtration is performed without adding extra sodium ions, as used in ion exchangers [43]. These membranes are the expensive part of the process and require to be changed timely.

Apart from radio-wastewater treatment, the use of nanofiltration has been extended to other industries also. Like, in Oil and Petroleum Chemistry, it is used for the removal of tar components and as a purifier of gas condensates.

In pharmaceuticals, they are used for room temperature solvent exchange and managements. Also, researchers can do extraction of amino acids and lipids from blood and other cell cultures using this nanofiltration technique.

OTHER METHODS HELPFUL IN RADIOWASTE WATER CLEANUP - PLANT -REMEDIATION

Plants helpful in radio-wastewater remediation, *i.e.* treatment of wastewater with the help of plants is called plant remediation or Phyto-remediation. For example, *Agrobacterium rhizogenes*, a bacteria found in the roots of some plants can significantly increase radionuclide uptake from the soil and water. These plant species help to remove radio-waste from the environment without suffering toxicity. They may lock these toxic wastes in certain areas of their roots or on the ground surface or make them volatile.

Rhizofiltration

Bioremediation technologies are one of the successful processes in the remediation of Cesium-137 and Strontium-90. These bio-redmedies use algae and aquatic plants. The plants such as *Cladophora* can treat Cesium-137 and *Elodea* genera treats strontium-90 in wetland areas. These plant roots adsorb the radionuclides as precipitates. The sunflowers can remove around 90% of uranium from radiowaste-water by rhizo-filtration technique in a single day. Hence, apart from nanomaterials, these plants in nature are also helpful in the removal of heavy metal ions from water.

Phytosequesteration

Phyto-sequesteration is a self-technique of plants to extract radionuclides. This is another phytoextraction method where waste from the root system reaches the vascular tissue. The mass of such shoots of these plants becomes highly concentrated of radionuclides. Here, a low concentration of radioactive material present in soil and deep underground water can be removed. This process of plants is also known as phytoaccumulation or phyto-absorption process and is helpful to attract Cesium[137], Strontium[90], Technetium[99], Cerium[144], Plutonium[240], Americium[241], Neptunium[237] and various radioisotopes of Thorium and Radium [44] for example, mustard greens and bok choy can accumulate Cesium and Uranium.

Phytovolatilization

Generally, radionuclides are less toxic in their volatile form in nature. Some plants are capable of capturing these radionuclides contaminants and transpire them later on (Fig. **7**).

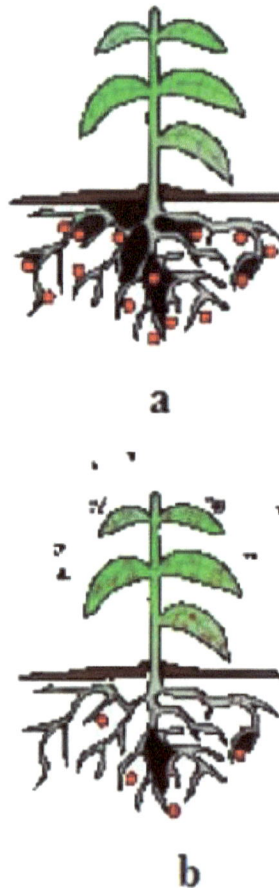

Fig. (7). a) Showing heavy and radioactive metals collected on roots and; **b)** the volatile contaminants are broken down and released into the atmosphere through plant leaves.

CONCLUSION

Nanomaterials due to their small volume and more surface area 1) become stronger, more stable and durable; and 2) materials may change electrical, optical, physical, chemical, or biological properties at the nano level and all these properties make chemical and biological reactions easier. This proved them the best decontaminant techniques for radiowaste-water purification.

Based on the results presented above, ferritin proved to have an excellent phosphate and Arsenic adsorption capacity. A high affinity for both oxyanions has been established underlining the potential of thermostable ferritin as a material for use in the production of clean drinking water and process water. In comparison with PRBs, nanoparticle injection allows remediation at greater depths and in areas unreachable by PRBs (*i.e.* land covered by a building). Apart from nanomaterials, plants having phytoremediation techniques are also useful for radiowaste-water remediation with a low operation cost. In future, new scientific discoveries and economic optimization in the field of nanotechnology for radio-wastewater treatment will benefit society.

CONSENT FOR PUBLICATION

Not applicable.

CONFLICT OF INTEREST

The authors declare no conflict of interest, financial or otherwise.

ACKNOWLEDGMENTS

I acknowledge all those who inspired me in writing this chapter. I also approve all the e-content available on the website, journals, books, *etc.,* making this chapter beneficial and valuable to the readers.

REFERENCES

[1]   Zhang W. Nanotechnology for water purification and waste treatment.Frontiers in Nanotechnology. Washington, DC: U.S. EPA Millennium Lecture Series 2005.

[2]   Cloete TE, de Kwaadsteniet M, Botes M, Lopez-Romero JM. Nanotechnology in water treatment applications. Norfolk: Caister Academic Press 2010.

[3]   Kunze C, Schulz R, Teunchens L, Velzen LV. Environmental remediation of radioactively contaminated sites, Environmental Radiation Survey and Site Execution Manual. EURSSEM 2010.

[4]   Yang D, Sarina S, Zhu H, *et al.* Capture of radioactive cesium and iodide ions from water by using titanate nanofibers and nanotubes. International Edition. Angewandte Chemie 2011; pp. 10594-8.

[5]   Song W, Grassian VH, Larsen SC. High yield method for nanocrystalline zeolite synthesis. Chem Commun (Camb) 2005; 20(23): 2951-3.

[http://dx.doi.org/10.1039/b501768h] [PMID: 15957036]

[6]   Song Weigu, Li Gonghu, Grassian Vicki H. Sarah C. Larsen. Development of Improved Materials for Environmental Applications: Nanocrystalline NaY Zeolites. Environmental Science and Techology 2005; 39(5): 1214-20.

[7]   Tratnyek PG, Johnson RL. Nanotechnologies for environmental cleanup. Nano Today 2006; 1(2): 44-8.
[http://dx.doi.org/10.1016/S1748-0132(06)70048-2]

[8]   Uyttebroek M, Baillieul H, Vermeiren N, *et al. In situ* remediation of a chlorinated ethene contaminated source zone by injection of zero-valent iron: from lab to field scale. Proceedings of PRB/RZ.

[9]   Li Z, Greden K, Alvarez PJJ, Gregory KB, Lowry GV. Adsorbed polymer and NOM limits adhesion and toxicity of nano scale zerovalent iron to E. coli. Environ Sci Technol 2010; 44(9): 3462-7.
[http://dx.doi.org/10.1021/es9031198] [PMID: 20355703]

[10]   Boronina T, Klabunde KJ, Sergeev G. Destruction of organohalides in water using metal particles: carbon tetrachloride/water reactions with magnesium, tin, and zinc. Environ Sci Technol 1995; 29(6): 1511-7.
[http://dx.doi.org/10.1021/es00006a012] [PMID: 22276871]

[11]   O'Carroll D, Sleep B, Krol M, Boparai H, Kocur C. Nanoscale zero valent iron and bimetallic particles for contaminated site remediation Adv Water Res 2013; 51: 104-22.
[http://dx.doi.org/10.1016/j.advwatres.2012.02.005]

[12]   Lien HL, Zhang W. Nanoscale iron particles for complete reduction of chlorinated ethenes. Colloids Surf A Physicochem Eng Asp 2001; 191(1-2): 97-105.
[http://dx.doi.org/10.1016/S0927-7757(01)00767-1]

[13]   Mohan CU. Pittman Jr. Arsenic removal from water/wastewater using adsorbents – a critical review. J Hazard Mater 2007. 142 (1–21-53).

[14]   Matias PM, Tatur J, Carrondo MA, Hagen WR. Crystallization and preliminary X-ray characterization of a ferritin from the hyperthermophilic archaeon and anaerobe *Pyrococcus furiosus*. Acta Crystallogr Sect F Struct Biol Cryst Commun 2005; 61(5): 503-6.
[http://dx.doi.org/10.1107/S1744309105011516] [PMID: 16511080]

[15]   Harrison PM, Arosio P. The ferritins: molecular properties, iron storage function and cellular regulation. Biochim Biophys Acta Bioenerg 1996; 1275(3): 161-203.
[http://dx.doi.org/10.1016/0005-2728(96)00022-9] [PMID: 8695634]

[16]   Honarmand Ebrahimi K, Hagedoorn PL, Hagen WR. Inhibition and stimulation of formation of the ferroxidase center and the iron core in *Pyrococcus furiosus* ferritin. J Biol Inorg Chem 2010; 15(8): 1243-53.
[http://dx.doi.org/10.1007/s00775-010-0682-6] [PMID: 20582559]

[17]   U.S. EPA. (2004) Cleaning Up the Nations Waste Sites: Markets and Technology Trends. EPA 542---04-015. Technology Innovation and Field Services Division. Office of Solid Waste and Emergency Response. http://www.epa.gov/tio/download/market/2004market.pdf

[18]   Gorman-Lewis D, Fein JB, Burns PC, Szymanowski JES, Converse J. Solubility measurements of the uranyl oxide hydrate phases metaschoepite, compreignacite, Na-compreignacite, becquerelite, and clarkeite. J Chem Thermodyn 2008; 40(6): 980-90.
[http://dx.doi.org/10.1016/j.jct.2008.02.006]

[19]   Yamamura T, Kitamura A, Fukui A, Nishikawa S, Yamamoto T, Moriyama H. Solubility of U(VI) in highly basic solutions. Radiochim Acta 1998; 83(3): 139-46.
[http://dx.doi.org/10.1524/ract.1998.83.3.139]

[20]   Silva RJ, Nitsche H. Actinide environmental chemistry. Radiochim Acta 1995; 70-71(s1): 377-96.
[http://dx.doi.org/10.1524/ract.1995.7071.s1.377]

[21]   Walther C, Denecke MA. Actinide colloids and particles of environmental concern. Chem Rev 2013; 113(2): 995-1015.
       [http://dx.doi.org/10.1021/cr300343c] [PMID: 23320457]

[22]   Parry SA, O'Brien L, Fellerman AS, *et al.* Plutonium behaviour in nuclear fuel storage pond effluents. Energy Environ Sci 2011; 4(4): 1457-64.
       [http://dx.doi.org/10.1039/c0ee00390e]

[23]   Vellora Thekkae Padil V, Rouha M, Černík M. Hydrocolloid-Stabilized Magnetite for Efficient Removal of Radioactive Phosphates. BioMed Res Int 2014; 2014: 1-10.
       [http://dx.doi.org/10.1155/2014/504760]

[24]   Tan XL, Xu D, Chen CL, Wang XK, Hu WP. Adsorption and kinetic desorption study of $^{152-154}$ Eu(III) on multiwall carbon nanotubes from aqueous solution by using chelating resin and XPS methods. Radiochim Acta 2008; 96(1): 23-9.
       [http://dx.doi.org/10.1524/ract.2008.1457]

[25]   Wang X, Chen C, Hu W, Ding A, Xu D, Zhou X. Sorption of 243Am(III) to multiwall carbon nanotubes. Environ Sci Technol 2005; 39(8): 2856-60.
       [http://dx.doi.org/10.1021/es048287d] [PMID: 15884386]

[26]   Chen C, Li X, Zhao D, Tan X, Wang X. Adsorption kinetic, thermodynamic and desorption studies of Th(IV) on oxidized multi-wall carbon nanotubes. Colloids Surf A Physicochem Eng Asp 2007; 302(1-3): 449-54.
       [http://dx.doi.org/10.1016/j.colsurfa.2007.03.007]

[27]   Chen CL, Li XL, Wang XK. Application of oxidized multi-wall carbon nanotubes for Th(IV) adsorption. Radiochim Acta 2007; 95(5): 261-6.
       [http://dx.doi.org/10.1524/ract.2007.95.5.261]

[28]   Balasubramanian K, Burghard M. Chemically functionalized carbon nanotubes. Small 2005; 1(2): 180-92.
       [http://dx.doi.org/10.1002/smll.200400118] [PMID: 17193428]

[29]   Kahn MGC, Banerjee S, Wong SS. Solubilization of oxidized single-walled carbon nanotubes in organic and aqueous solvents through organic derivatization. Nano Lett 2002; 2(11): 1215-8.
       [http://dx.doi.org/10.1021/nl025755d]

[30]   Schierz A, Zänker H. Aqueous suspensions of carbon nanotubes: Surface oxidation, colloidal stability and uranium sorption. Environ Pollut 2009; 157(4): 1088-94.
       [http://dx.doi.org/10.1016/j.envpol.2008.09.045] [PMID: 19010575]

[31]   Freeman WA, Mock WL, Shih NY. Cucurbituril. J Am Chem Soc 1981; 103(24): 7367-8.
       [http://dx.doi.org/10.1021/ja00414a070]

[32]   Lagona J, Mukhopadhyay P, Chakrabarti S, Isaacs L. The cucurbit[n]uril family. Angew Chem Int Ed 2005; 44(31): 4844-70.
       [http://dx.doi.org/10.1002/anie.200460675] [PMID: 16052668]

[33]   Sundararajan M. Designing novel nanomaterials through functionalization of carbon nanotubes with Supramolecules for application in nuclear waste management. Sep Sci Technol 2013; 48(16): 2391-6.
       [http://dx.doi.org/10.1080/01496395.2013.807829]

[34]   Shokouhimehr M, Soehnlen ES, Khitrin A, Basu S, Huang SD. Biocompatible Prussian blue nanoparticles: Preparation, stability, cytotoxicity, and potential use as an MRI contrast agent. Inorg Chem Commun 2010; 13(1): 58-61.
       [http://dx.doi.org/10.1016/j.inoche.2009.10.015]

[35]   Namiki Y, Namiki T, Ishii Y, *et al.* Inorganic-organic magnetic nanocomposites for use in preventive medicine: a rapid and reliable elimination system for cesium. Pharm Res 2012; 29(5): 1404-18.
       [http://dx.doi.org/10.1007/s11095-011-0628-x] [PMID: 22146802]

[36]　Rosoff B, Cohn SH, Spencer H. Cesium-137 metabolism in man. Radiat Res 1963; 19(4): 643-54.
[http://dx.doi.org/10.2307/3571485] [PMID: 14062237]

[37]　Thammawong C, Opaprakasit P, Tangboriboonrat P, Sreearunothai P. Prussian blue-coated magnetic nanoparticles for removal of cesium from contaminated environment. J Nanopart Res 2013; 15(6): 1689-98.
[http://dx.doi.org/10.1007/s11051-013-1689-z]

[38]　Dresow B, Nielsen P, Fischer R, Pfau AA, Heinrich HH. In vivo binding of radiocesium by two forms of Prussian blue and by ammonium iron hexacyanoferrate (II). J Toxicol Clin Toxicol 1993; 31(4): 563-9.
[http://dx.doi.org/10.3109/15563659309025761] [PMID: 8254698]

[39]　Faustino PJ, Yang Y, Progar JJ, *et al.* Quantitative determination of cesium binding to ferric hexacyanoferrate: Prussian blue. J Pharm Biomed Anal 2008; 47(1): 114-25.
[http://dx.doi.org/10.1016/j.jpba.2007.11.049] [PMID: 18242038]

[40]　Verzijl JM, Joore JCA, van Dijk A, *et al. In Vitro* Binding characteristics for cesium of two qualities of prussian blue, activated charcoal and Resonium-A. J Toxicol Clin Toxicol 1992; 30(2): 215-22.
[http://dx.doi.org/10.3109/15563659209038633] [PMID: 1588671]

[41]　Van der Bruggen B, Mänttäri M, Nyström M. Drawbacks of applying nanofiltration and how to avoid them: A review. Separ Purif Tech 2008; 63(2): 251-63.
[http://dx.doi.org/10.1016/j.seppur.2008.05.010]

[42]　Hilal N, Al-Zoubi H, Darwish NA, Mohamma AW, Abu Arabi M. A comprehensive review of nanofiltration membranes:Treatment, pretreatment, modelling, and atomic force microscopy. Desalination 2004; 170(3): 281-308.
[http://dx.doi.org/10.1016/j.desal.2004.01.007]

[43]　Baker LA. Current Nanoscience. Nanomedicine 2006; 2(3): 243-55.

[44]　Dushenkov S. Trends in phytoremediation of radionuclides. Plant and Soil Netherlands 2003; 249(1): 167-75.
[http://dx.doi.org/10.1023/A:1022527207359]

CHAPTER 8

Nanomaterials in Organic Synthesis

Shraddha Tivari[1,*], **Manoj Kumar**[1], **Seraj Ahmad**[1], **Akram Ali**[1] and **Vishal Srivastava**[1]

[1] *Department of Chemistry, CMP Degree College, University of Allahabad, Uttar Pradesh 211002, India*

Abstract: Multidisciplinary research in chemistry, physics and other engineering sciences often addresses nanotechnology. In almost all branches of science and technology, nanotechnology is commonly used. Nanomaterials are not just something developed in the laboratory but nanotechnology has made it possible for humans to manufacture nanoform-containing materials. Metal nanoparticles have been used in different areas such as catalysis, sensor, and medicine. Nanoparticles have good efficiency, selectivity and yield of catalytic processes. Nanoparticles have higher selectivity in the reactions because the reactions continue with fewer impurities and less waste. Hence this technique is safer and more environmental-friendly. The specific emphasis of this chapter is on the applications of nanoparticles in organic synthesis.

Keywords: Green Synthetic Method, Dichroism, Surface Effect, Nanotechnology, Nanoparticles, Mizoroki-Heck Reactions.

INTRODUCTION

In scientific literature, the term nano derived from the Greek "nanos" which means "dwarf", is becoming increasingly popular. Nanoparticles are materials ranging in size between 1 to 100 nm in a single unit. There are many examples of nanomaterials such as titanium dioxide, silver, synthetic amorphous silica, iron oxide, azo pigments and Phthalocyanine pigments. One of the most fascinating examples of nanotechnology from the ancient world is the Roman Lycurgus cup. Its dichroic glass changes hue, which means that the cup has two different colors. When in light, the dichroic glass appears red-purple in color.

The scientists studied the cup in 1990 using transmission electron microscopy (TEM) that was used to clarify the dichroism phenomenon. The dichroism phenomenon is attributable to the existence of nanoparticles with a diameter of 50-100 nm. These nanoparticles have been found to be silver-gold alloys, which

* **Corresponding author Shraddha Tivari:** Department of Chemistry, CMP Degree College, University of Allahabad, Uttar Pradesh 211002, India; E-mail: tripathishraddha934@gmail.com

Arti Srivastava, Mridula Tripathi, Kalpana Awasthi and Subhash Banerjee (Eds.)

have a 7:3 ratio and contain around 10 percent copper (Cu) in a glass matrix which is dispersed. Au nanoparticles absorb light (~520 nm), and produce red color. The red-purple color is due to absorption by the bigger particles while the green color is attributable to the dispersion light by colloidal dispersion of Ag nanoparticles > 40nm in size. Nanotechnology, now has become a more academic focus. This technology is being used to treat several illnesses. Nanotechnology can be described as "The nanoscale."

NANOMATERIALS VS BULK MATERIALS

There are mainly two primary factors which distinguish nanomaterials from bulk materials; one is surface effects (causing smooth scaling properties due to surface fraction atoms) and quantum effects (showing discontinuous behavior due to quantum confinement effect in materials with delocalization of electrons). Such variables influence their magnetic, optical, electrical and mechanical properties.

IMPORTANCE OF NANOMATERIALS

Nanomaterials are widely distributed and are very valuable for all kinds of practical uses as shown in Fig. (**1**). These materials are not just something made in the laboratory but nanotechnology has made it possible for humans to create materials that include nano form.

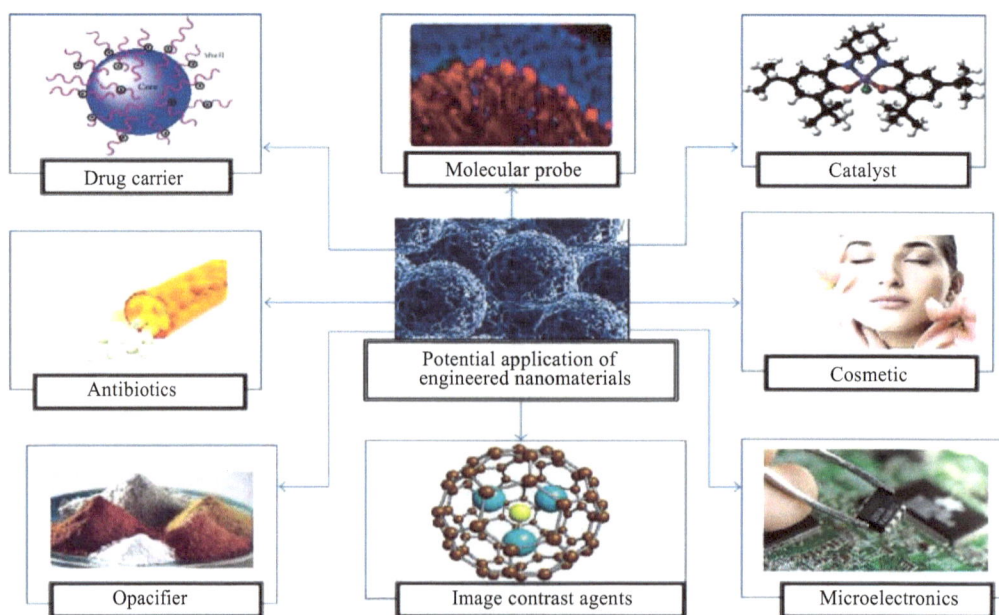

Fig. (1). Applications of Nanomateriales.

APPLICATIONS OF NANOPARTICLES IN ORGANIC SYNTHESIS

Nanomaterials have often been used in textiles, nanofibers, nanowires, coatings and plastics. Nanomaterials in reactions have higher selectivity. There are less waste and less impurity in these reactions, hence this is a safer technique and more environmental-friendly.

Iron nanoparticles are used in different organic synthesis reactions. These nanoparticles are synthesized as green solvent using polymers taken as a capping agent in water. The value of iron nanoparticles is that hydrogen peroxide is catalyzed for the treatment of organic pollution and also used as environmentally benign catalysts for alkenes and alkynes hydrogenation.

Zhang Z-H *et al.* described the formation of tetraalkylpyrazine with the presence of Fe_3O_4 and various solvent like acetonitrile, toluene, dichloromethane, ethyl acetate, ethanol and water (Scheme **1**). The solvent has been shown to play an important role in product yield. Generally non-polar solvent such as ethyl acetate, toluene and dichloromethane afforded low yield of product. In the presence of water, the best conversion takes place. The product yield was found to be influenced by the amount of Fe_3O_4 and the best result was obtained when 10 mol-percent Fe_3O_4 was used during the reaction in water at room temperature [1].

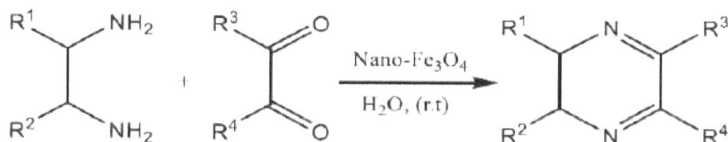

Scheme 1. Synthesis of tetraalkylpyrazine.

In the presence of Pt-metal complexes, silane was hydrolytically oxidized. In the above mentioned reaction, Pt-nanoparticles provide a selective route to the formation of silanols (Schemes **2** & **3**) [2 - 4].

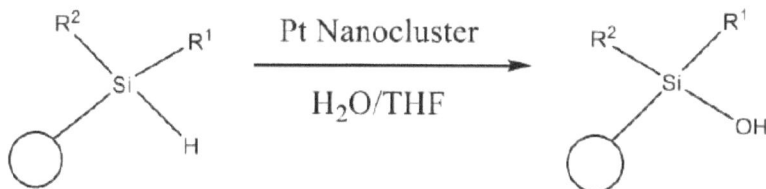

Scheme 2. Formation of Silanols.

$$\text{PhMe}_2\text{Si-H} \xrightarrow{\text{Pd/AuNp's}} \text{PhMe}_2\text{Si-OH}$$

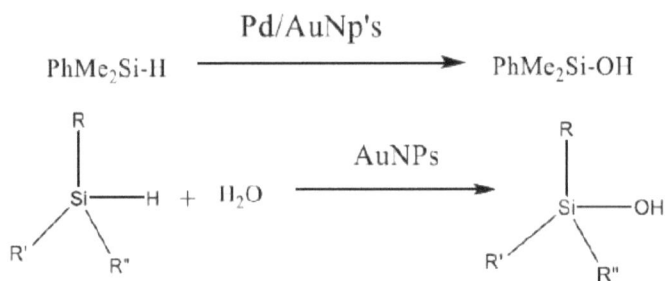

Scheme 3. Synthesis of silanols.

Due to its size and morphological dependent optical, physical, electrical and chemical properties, gold nanoparticles have gained more research attention in recent years. Due to the vivid colors created by their interaction with visible light, gold nanoparticles are used by artists. Another significant application is for colorimetric identification of heavy toxic metals such as Hg^{2+}, Cd^{2+}, As^{3+}, Pb^{2+} *etc.*, since gold nanoparticles have strong selectivity and sensitivity. In the absence of organic ligands Au-Pd nanomaterials are synthesized and absorbed in support of TiO_2 and are found to be stable under oxidative catalysis conditions [5].

Aromatic azo compounds have many characterstics so these compounds are used in many areas such as food additives, pharmaceuticals products and dye processing. These compounds have recently been synthesized under high temperatures and high pressures using transition metals as reducing agents. The by-products derived from reducing agents are not environment friendly. Currently, azo-aromatic compounds have been reported to be synthesized by using corresponding nitro aromatic compounds through a two-step, one-pot catalyst reaction consisting of AuNPs on TiO_2 or CeO_2 at 100°C or higher (Scheme **4**) [6].

1-nitrobenzene $\xrightarrow{\text{Au/ZrO}_2}$ 1,2-diphenyldiazene   + O_2

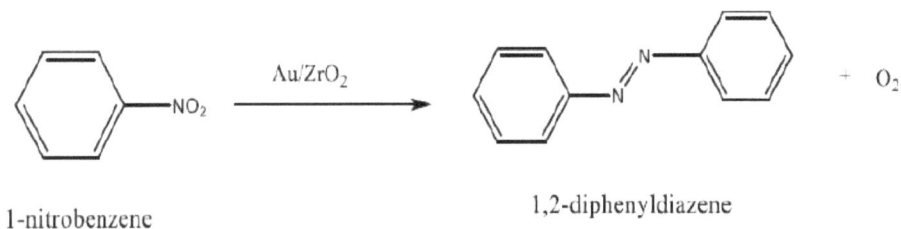

Scheme 4. Synthesis of 1, 2-diphenyldiazine from nitrobenzene.

Ni is often found free in nature but it is usually found in ores. It can be alloyed with iron, tungsten, molybdenum and with other metals for the formation of corrosion resistant alloys. Ni nanoparticles are used for many applications such as magnetic fluid, propellant, sintering additive, coating, nanowires, nanofibres, automotive catalytic converters and catalysts.

Ni nanoparticles are embedded in silica and carbon aerogel. These nanoparticles are recoverable and active catalysts for Mizoroki-Heck reactions. Peng *et al.* described the above scheme in which Ni nanoparticles were used as catalysts in hydrothermal Heck reaction (Scheme 5). Ni nanoparticles were prepared by direct reduction of $NiCl_2$ with $NaBH_4$ in aqueous medium under sonication in the absence of surfactants. Under optimum reaction conditions, the coupling reactions between a variety of aryl halides and different alkenes were studied. The inexpensive catalysts have maintained good to moderate activity and are stable in the reaction conditions [7].

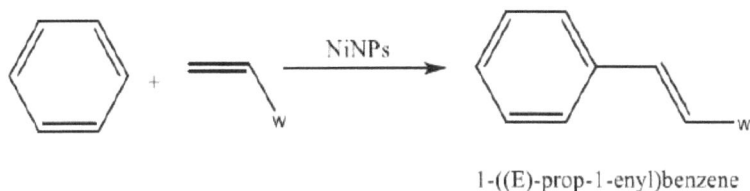

1-((E)-prop-1-enyl)benzene

Scheme 5. Synthesis of 1-((E)-prop-1-enyl) benzene.

Ag nanoparticles were synthesized by using a photochemical green synthetic method. Ag nanoparticles were synthesized using the method of seed-mediated growth and even in presence of ionic liquids [8]. Silver nanoparticles have various applications in food storage bags, cosmetic products and the textiles industry. For the reduction reaction of 4-nitrophenol, these nanoparticles are used as a catalyst.

The nitrile-amide hydration reaction in water is efficiently catalyzed by silver nanoparticles. The Ag catalyst can be recycled under neutral conditions in this process and water is taken as solvent. This method is safer for the ecosystem and more rewarding for the industry. Silver nanoparticles may serve as heterogeneous catalysts in this method. Ag nanoparticles were centrifuged and were able to be reused without any loss of catalytic activity. This method has been applied for the synthesis of several amides with good yields (Scheme 6) [9].

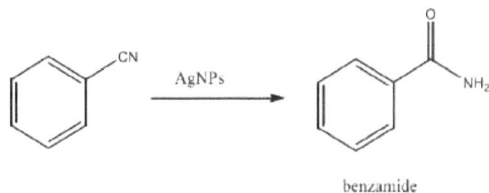

benzamide

Scheme 6. Synthesis of benzamide.

CONCLUSION

For chemical industry, nanomaterials are very important for many processes such as purification of effluents, industrial waste gases and for catalysts. Catalysts have been used to work with fuels like gas, coal *etc*. Recently heterogeneous catalysts have gained a lot of popularity in comparison to homogeneous catalysts because they show more selectivity and better yield of products in reactions. In order to enhance the catalytic process, nanomaterials that are used as catalysts are typically heterogeneous catalysts broken up into metal nanoparticles. Catalysts of Metal nanoparticles can be easily isolated and recycled. These nanoparticle catalysts have the potential to increase catalytic activity because they have high surface area. Metal nanoparticles for several green reactions have been used as catalysts. In these reactions, water was used as solvent and ionic liquids were also used.

CONSENT FOR PUBLICATION

Not applicable.

CONFLICT OF INTEREST

The authors declare no conflict of interest, financial or otherwise.

ACKNOWLEDGEMENT

Declared none.

REFERENCES

[1]   Lü HY, Yang SH, Deng J, Zhang ZH. Magnetic Fe_3O_4 Nanoparticles as New, Efficient, and Reusable Catalysts for the Synthesis of Quinoxalines in Water. Aust J Chem 2010; 63(8): 1290-6.
[http://dx.doi.org/10.1071/CH09532]

[2]   Chauhan BPS, Sarkar A, Chauhan M, Roka A. Water as green oxidant: a highly selective conversion of organosilanes to silanols with water. Appl Organ met. Chem 2009; 23: 385-90.

[3]   Jeon M, Han J, Park J. Transformation of silanes into silanols using water and Recyclable Metal Nanoparticle Catalysts. ChemCatChem 2012; 4(4): 521-4.
[http://dx.doi.org/10.1002/cctc.201100456]

[4]   Gitis V, Beerthuis R, Shiju NR, Rothenberg G. Organosilane oxidation by water catalysed by large gold nanoparticles in a membrane reactor. Catal Sci Technol 2014; 4(7): 2156-60.
[http://dx.doi.org/10.1039/C3CY00506B]

[5]   Frank AJ, Rawski J, Maly KE, Kitaev V. Environmentally benign aqueous oxidative catalysis using $AuPd/TiO_2$ colloidal nanoparticle system stabilized in absence of organic ligands. Green Chem 2010; 12(9): 1615-22.
[http://dx.doi.org/10.1039/c0gc00084a]

[6]   Zhu H, Ke X, Yang X, Sarina S, Liu H. Reduction of nitroaromatic compounds on supported gold nanoparticles by visible and ultraviolet light. Angew Chem Int Ed 2010; 49(50): 9657-61.
[http://dx.doi.org/10.1002/anie.201003908] [PMID: 21053223]

[7]   Zhang W, Qi H, Li L, *et al.* Hydrothermal Heck reaction catalyzed by Ni nanoparticles. Green Chem 2009; 11(8): 1194-200.
[http://dx.doi.org/10.1039/b900697d]

[8]   Kim AY, Bae HS, Park S, Park S, Park KH. Silver Nanoparticle Catalyzed Selective Hydration of Nitriles to Amides in Water Under Neutral Conditions. Catal Lett 2011; 141(5): 685-90.
[http://dx.doi.org/10.1007/s10562-011-0561-y]

[9]   Quaresma P, Soares L, Contar L, *et al.* Green photocatalytic synthesis of stable Au and Ag nanoparticles. Green Chem 2009; 11(11): 1889-93.
[http://dx.doi.org/10.1039/b917203n]

Part 3: Functional Materials for Energy Conservation

Implications and Applications of Multifunctional Advanced Materials/Gadgets for Energy Conversion and Storage

Pinki[1], **Subhash**[1] and **Ashu Chaudhary**[1,*]

[1] *Department of Chemistry, Kurukshetra University, Kurukshetra, Haryana, India*

Abstract: The expanded interest in vitality assets, extraordinary endeavors, advocacy of convenient hardware and electric vehicles globally animates the improvement of energy storage gadgets, *e.g.,* lithium-ion batteries and supercapacitors, toward higher energy density, which essentially relies on new materials utilized in these gadgets. Besides, energy storage materials assume a key part in productive, clean, and adaptable utilization of energy, and are vital for exploiting sustainable power systems. The usage of the thermal energy storage (TES) framework with phase change material (PCM) is a viable route for energy preservation and green-house gas emission reduction. Ongoing advances in atomically thin two-dimensional transition metal dichalcogenides (2D-TMDs) have prompted an assortment of promising innovations for nanoelectronics, photonics, energy storage, and so on. Graphene and graphene-based materials have attracted extraordinary consideration due to their interesting properties of high mechanical adaptability, huge surface zone, chemical stability, prevalent electric and thermal conductivities that render them incredible as alternative electrode materials for electrochemical energy storage frameworks. The straightforward Chemical Vapour Deposition (CVD) and Atomic Layer Deposition (ALD) approaches offer another route for the creation of permeable materials for energy storage. Alteration of organic substrates with inorganic polyoxometalate (POM) clusters can be utilized to build nanocomposite materials with improved properties and various functionalities. Nanotechnology offers up new frontiers in materials research and construction to address the energy challenge by forming novel materials, particularly carbon nanoparticles, for efficient energy transformation and capacity, Polyaniline (PANi) as an auspicious material for energy storage/transformation, is merited for serious investigation and further progress. This book chapter discusses the various methods in materials for energy, their storage, and applications in numerous fields.

* **Corresponding author Ashu Chaudhary:** Department of Chemistry, Kurukshetra University, Kurukshetra, Haryana, India; Pin-136119, Tel: +91- 9729864551; E-mail: ashuchaudhary21@gmail.com

Arti Srivastava, Mridula Tripathi, Kalpana Awasthi and Subhash Banerjee (Eds.)

Keywords: Energy Conversion & Storage, 2D-TMDs, Polyoxometalates, Polyaniline, Supercapacitors, Fuel Cells.

INTRODUCTION

The prerequisite for elective and sustainable sources of power turns out to be more critical as we move further into the 21st century [1]. Spotless, modest, and safe high energy and power density resources are needed to empower the generation, stockpiling, and transformation of energy through non-contaminating methods. Energy storage is cultivated by gadgets or physical media that store energy to perform valuable activities. The energy present at the underlying development of the universe has been put away in stars, for example, the sun, and is currently being utilized by humans legitimately (for example through sun-oriented warming), or in a roundabout way (for example by growing yields or transformation into power in solar-based cells). Researchers have never given up trying to improve the electrochemical presentation of existing energy storage device measures in order to satisfy the need of individuals who yearn for a reliable and clean power source that can keep up with advancements in daily life and invention [2].

Among the innovations for not-so-distant future energy stockpiling are battery-powered lithium-ion batteries, fuel cells, and supercapacitors (SCs) [3]. In the previous twenty years, lithium-ion batteries (LIBs) have accomplished extraordinary triumphs because of their superior exhaustive battery execution, contrasted with lead-acid, Ni-Cd and Ni-MH batteries. In any case, similar to the past battery procedures, its advancement has likewise met a bottleneck in the present time [4]. Supercapacitors (SCs) overcome any barrier between ordinary capacitors and battery-operated types of equipments. Fuel cells (FCs) are a gadget that renovates over the chemical energy from a fuel into power through a chemical reaction with oxygen or another oxidizing agent.

Around the world, there is an expanding research enthusiasm on solar-powered thermal energy as plentiful, modest, viable, and clean energy. However, a never-ending energy source using solar thermal energy is being searched due to the absence of daylight at night or its rarity on a gloomy day. In that unique circumstances, researchers are searching for an approach to store this huge energy to use it without daylight/sunlight. It could be the arrangement of the two present issues, for example, the investigated consumption of petroleum derivatives at a high rate and the natural impacts of global warming [5].

Thermal energy storage (TES) framework with phase change material (PCM) could be a decent alternative to decrease these issues. It is likewise required to control the present hazardous global warming rate. Over the most recent 150 years, extremist advancement of industrialization and related human exercises have collected tremendous measures of GHGs to the environment [6]. Alexiadis (2007) arranged a model to measure the impact of the CO_2 outflow because of anthropogenic exercises. He indicated that the anthropogenic CO_2 discharge has become the principal driving force in an unnatural weather change [7]. Ozone-depleting substances from the burning of petroleum derivatives, production, transportation, and energy transformation bring about atmosphere changes by influencing the environment artificially in the long term [8].

An energy stockpiling measure takes a shot at three principal exercises, for example, charging (stacking), putting away, and releasing (delivering) [9]. In addition, one of the flourishing methods to store thermal energy is the phase change material execution. Thermal energy storage frameworks can assist with upgrading energy productivity and relieve energy-related ecological impacts particularly in building heating, cooling, and power generation. Subsequently, thermal energy storage frameworks could assume an essential part for climate by moderating emissions of CO_2, SO_2, NO_x, and CFCs [10]. The incredible accomplishment of graphene has been trailed by an equally impressive surge in the advancement of other 2D materials that can frame nuclear sheets with exceptional properties. These incorporate 2D transition metal dichalcogenides (for example molybdenum disulfide (MoS_2), molybdenum diselenide ($MoSe_2$), tungsten disulfide (WS_2), and tungsten diselenide (WSe_2)), hexagonal boron nitride (h-BN), borophene (2D boron), silicene (2D silicon), germanene (2D germanium), and MXenes (2D carbides/nitrides) [11 - 18].

As a building block for carbon constituents of every other dimensionality, the two-dimensional (2D) single nuclear carbon sheet of graphene has immediately developed as an alluring contender for energy applications because of its interesting structure and properties, including a huge explicit surface territory, fantastic electrical conductivity, great warm conductivity, and high charge versatility, extraordinary mechanical quality, low optical absorbance (2.3%), and density and eccentric adaptability [19, 20]. Like every other material, however, it is extremely uncommon for a graphene-based material with desirable bulk properties to likewise have the surface qualities required for certain particular applications. Consequently, surface functionalization is fundamental in making graphene materials of good bulk and surface properties as requested for effective energy transformation and capacity.

Polyoxometalates (POMs), at class of metal-oxygen groups of the early transition elements, are the absolute most important structural blocks for nanocomposite materials. Polyoxometalates have an unequaled scope of physical and chemical properties which emerge from their interminable assortment of atomic structures and proportions [21]. For example, polyoxometalates can cover the range from 6 to 368 metal particles in a solitary molecule, and the monodisperse size can be specifically adjusted from a few angstroms to 10 nm [22]. The assorted scope of functionalities showed by this great class of particles can be misused in various ways .

Besides, the antiviral activity of certain polyoxometalates has been utilized in clinical applications and their photochemical conduct has prodded enthusiasm for polyoxometalate-modified terminals for photovoltaics [21, 23]. These particles exhibit high dependability of their oxidation and reduction states and can take an interest in quick reversible electron transfer reactions [24]. These properties have led some authors to depict polyoxometalates as electron repositories or wipes [25], an ideal trademark for energy stockpiling applications. Regarding redox and electron transfer properties, polyoxometalates intently take after related transition metal oxides which have been utilized widely in energy stockpiling anodes for batteries and electrochemical capacitors (or supercapacitors) [26].

Among all the testimony measures, atomic layer deposition (ALD) [27, 28] seems, by all accounts, to be one of the most important procedures because of its straightforwardness, reproducibility, and high conformality of the obtained films. In actuality, atomic layer deposition was at that point picked by the microelectronic business as the method of decision for the manufacture of next-generation metal-insulator-metal capacitors in DRAMs and to deposit gate oxide in CMOS semiconductors [29]. High-proportion structures, nanoparticles, nanowires, nanotubes, delicate materials, and natural materials can be exactly covered by atomic layer deposition [27, 28, 30]. The obtained heterostructures have a great scope of utilizations, for example, in catalysis, microelectronics, energy stockpiling and conversion, detecting, and so forth [27, 28, 30].

The morphology of the polyaniline got from electrochemical testimony is typically restricted to nanofibers, or thin film at the surface of substrates. In addition, the morphologies of polyaniline are greatly subject to the properties of the substrates. PANi has been broadly utilized in energy stockpiling and conversion gadgets, including supercapacitors, batteries, and power modules. At the point when utilized for supercapacitors, PANi as the dynamic material stores charge through redox reactions as the PANi progresses between different oxidation states. It has had the option to accomplish explicit capacitance as high as 950 F g^{-1} through the inclusion of the whole volume in the storage of charge, outperforming other conducting polymers that store charge exclusively on the surface [31]. The chemical structure of polyaniline is shown in Fig. (**1**).

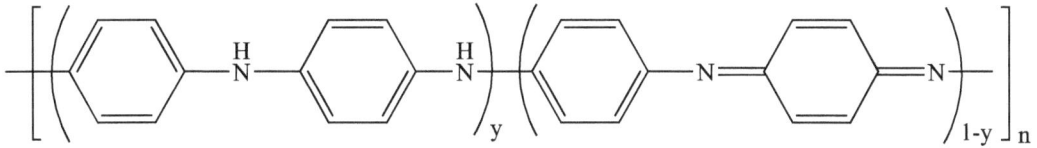

Fig. (1). The asymmetric chemical structure of PANi [32].

To meet up the requirement for the near-future energy storage device, high power, and energy density energy storage devices call for consideration. Fig. (**2**) shows the different types of energy storage and conversion devices. In this book chapter, we discuss energy storage and conversion devices.

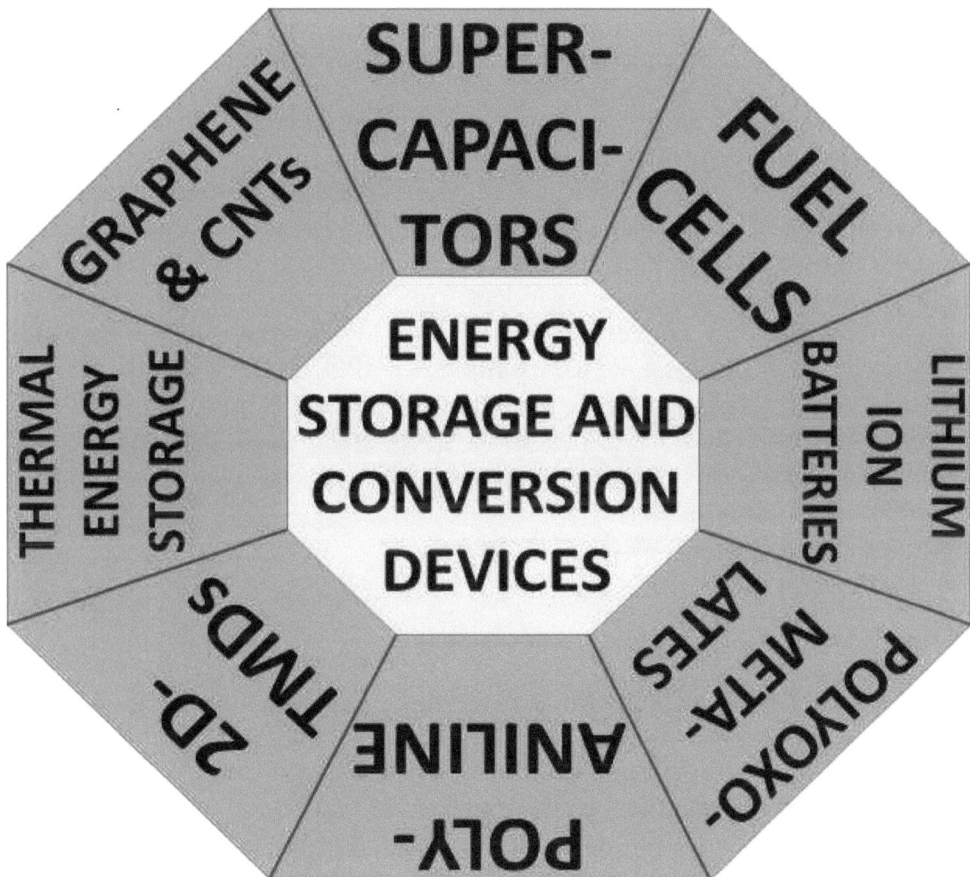

Fig. (2). Various energy storage and conversion devices.

VARIOUS METHODS FOR ENERGY STORAGE AND CONVERSION

Energy Storage and Conversion by Mithium-ion Batteries

Sometime in the past, lead-acid batteries were the main ones found in our vehicles and Ni-Cd was omnipresent in our camcorders. At that point, after a concise problem between the regnum of metal hydrides, lithium assumed control over the compact gadgets market [33]. In the recent past, that was the main predictable use of lithium-ion batteries, and research was motivated by the growing need for small devices with the primary goals of greater autonomy, energy density, and plan adaptability [34]. In under 10 years, attention has moved to quicker charging, longer-enduring, high-power, yet ecologically neighbourly Li-ion batteries which could likewise be utilized for electric vehicles in a supportable manner [35]. The rundown of standard materials concentrated to upgrade the exhibition of commercial, original lithium batteries is currently very lengthy [34].

An assortment of inorganic phases, including generally oxides and phosphates, have been studied as cathode materials ($Li_{1-x}Ni_{1-y-z}Co_yM_zO_4$, $Li_xMn_2O_4$, $LiMnO_2$, MnO_2, V_2O_5, LiV_3O_8, $Li_{1-x}VOPO_4$, Li_xFePO_4). Anode chemistries are much more assorted since they involve an assortment of materials as well as an assortment of approaches, from lithium metal to graphite and carbons to amalgams (Li with Si, Sn...) to the later anodes dependent on metal oxides prompting metal nanoparticles through conversion responses [34]. In the specific instance of ineffectively leading materials, for example, $LiFePO_4$, surface modifications with carbon have been set up as a fundamental advance for their usage as cathode dynamic materials and have prompted a momentous improvement of their presentation, particularly at high discharge rates, in this manner adding to the combination of batteries as energy stockpiling gadgets with high trusts in high-power and quick charging marks.

TES with PCM

The reasonable improvement must meet three fundamental things affordable, natural, and social [36]. In this perspective, thermal energy stockpiling is one of the planned energy stockpiling frameworks for energy protection. Thermal energy could be put away as chemical energy (reversible reactions), reasonable heat, and latent heat (LH). Thinking about high-energy density and small temperature deviation from capacity to recovery, latent heat is profoundly important [37]. Thermal energy storage frameworks store heat as latent heat for additional utilization. This is an incredible chance to store accessible sun-based energy to utilize later for application. The storage material assimilates heat during softening while the material holds its temperature fixed at the dissolving temperature.

Melting completes with engrossing the melting enthalpy and further heat movement creates reasonable heat stockpiling. This dissolving enthalpy/latent heat is utilized to store energy as warmness. Materials with a solid-liquid (melting) or solid-solid phase change which are appropriate for hot or cold stockpiling utilized as latent heat stockpiling material or essentially phase change material (PCM) [38]. The latent heat property of phase change materials is the most positive for application as it acts as a thermal switch. Arriving at the melting point, the capacity material proceeds with the consistent temperature for some time to be dissolved completely. This phase change measure empowers the ingestion of a lot of heat without expanding the temperature of the framework.

A phase change material can without much of a stretch be applied into a current thermal administration framework because the latent heat property of the phase change material is a characteristic cycle that needn't to be substituted with any additional energy contribution from the framework [39]. These phase change materials could be utilized for space warming, space cooling, power generation, green-house warming, sun-oriented cooking, waste heat recuperation system, and latent heat storage exchanger with the reasonable latent heat property [40]. Besides, phase change materials likewise could be applied to cool the lithium-ion battery which are utilized in electric vehicles [41]. Subsequently, applications of phase change materials may contribute to various segments for energy preservation. The flow chart for the thermal energy storage is shown in Fig. (**3**).

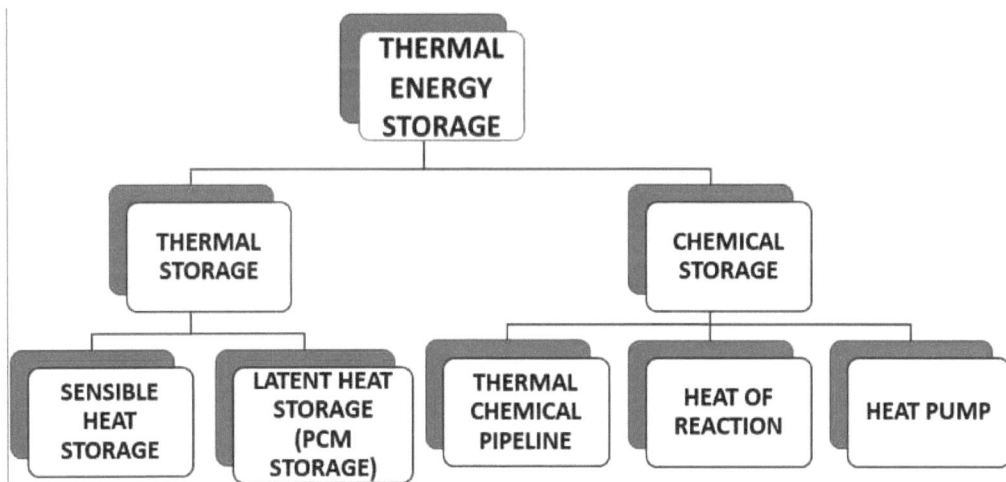

Fig. (3). Thermal energy storage and their types.

Supercapacitors

The cathodes for superconductors are the fundamental question mark. Carbon materials, transition metal oxides, and directing polymers have been generally examined as supercapacitor cathodes. Carbon-based materials, for example, carbon nanotubes, and graphene have noticeable electrochemical stability and high electrical conductivity, yet they experience the ill effects of low explicit capacitance (90 to 250 F/g), which make them exceptionally challenging to grow high energy density superconductors [42]. Then again, metal oxides and conducting polymers can convey considerably higher explicit capacitances of 300-1200 F/g through Faradaic reactions. In any case, their poor conductivity and general kinetic irreversibility restrict their viable application for high energy density super-conductors [43].

Two-dimensional Transition Metal Dichalcogenides

Significant endeavors have been committed to the synthesis of controllable, great scope, and uniform nuclear layers of different 2D transition metal dichalcogenides utilizing different top-down and bottom-up approaches, including mechanical shedding, substance peeling, and CVD. The greater part of the announced information and hypothesis on the essential material science and gadgets on two-dimensional transition metal dichalcogenides have generally depended on the peeling strategy because of its high caliber. Be that as it may, the basic constraints of the flake size and film consistency have hauled its advancement past the key examinations. Unexpectedly, the chemical vapor deposition cycle has been read for the adaptable and reliable production of great area two-dimensional transition metal dichalcogenides.

In any case, chemical vapor deposition developed transition metal dichalcogenides show low quality when contrasted with their peeled counterparts. Recently, endeavors have been made to acquire great transition metal dichalcogenides with thickness controllability and wafer-scale consistency utilizing atomic layer deposition (ALD), metal-organic CVD (MOCVD), and direct statement techniques (faltering, pulsed laser deposition (PLD), e⁻-beam). The two-dimensional materials shaping synthetic reactions by and large utilize either thermal energy from a warmed substrate or non-thermal energy, for example, microwave or photon energy into the response cycle, and the two-dimensional materials framing measure relies upon grid boundary of substrates, temperatures, and nuclear gas transition [44, 45].

The great interest in growing exceptionally sensitive, particularly, low power devouring, dependable and compact sensors has animated broad exploration of new detecting materials dependent on two-dimensional allotropes of transition metal dichalcogenides and phosphorous after the incredible achievement made by their two-dimensional carbon analog, that is, 'graphene'. The high surface-t--volume proportion in two-dimensional transition metal dichalcogenides offers colossal potential for the location of a lot of target experts per unit area just as quick reaction and improvement with low power utilization [46, 47]. In addition, the ongoing exhibition of the versatile union of two-dimensional transition metal dichalcogenides has demonstrated the possibility to manufacture financially cost-effective sensors. Fig. (**4**) shows the utilization of two-dimensional transition metal dichalcogenides, for example, MoS_2, WS_2, and so on for different detecting applications including gas, chemical, and bio-sensors (like DNA and glucose sensors).

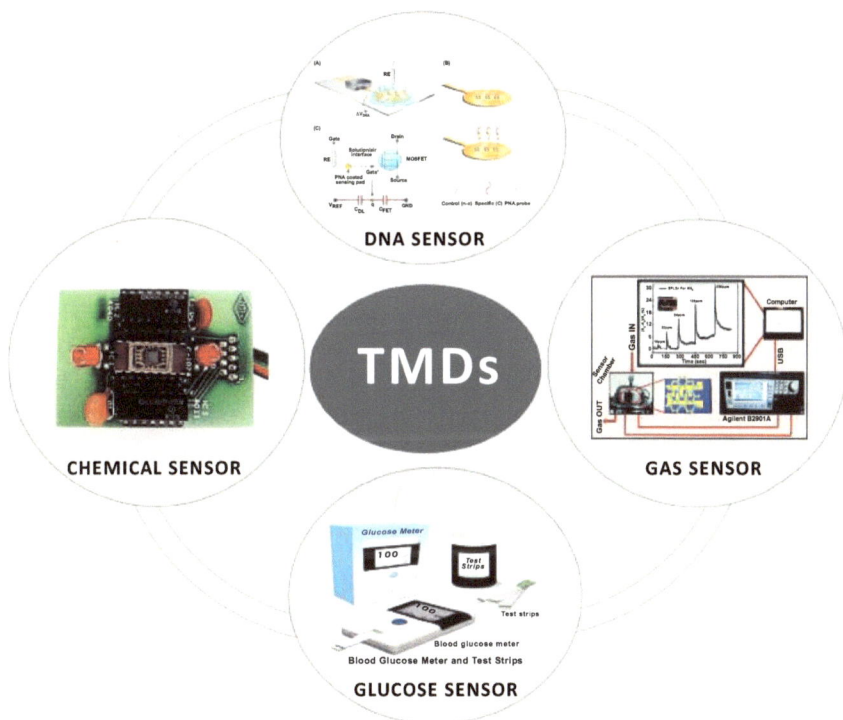

Fig. (4). Applications of two-dimensional transition metal dichalcogenides in different types of sensors.

True to form, a large portion of the detailed two-dimensional transition metal dichalcogenides-based sensors have been acknowledged utilizing precisely exfoliated or fluid stage exfoliated MoS_2 fragments [48, 49]. For instance, Li *et al*. [50] manufactured MoS_2-FET sensor gadgets utilizing mechanically exfoliated single and not many layer MoS_2 flicks to distinguish nitric oxide (NO). It was seen that MoS_2 films showed a high affectability to NO with a low location breaking point of 0.8 ppm. An artificially peeled MoS_2 fragments-based sensor by Donarelli *et al.* [51] out-performs comparative sensors, with a deliberate recognition limit of 20 ppb when presented to NO_2 gas.

The probable mechanism for the p-type conduct of the detecting material (MoS_2) toward NO_2 is the N substitutional doping of S opening in the MoS_2 surface. This identification limit is the most minimal ever estimated to nitrogen dioxide for MoS_2-based sensors and is similar or better than ZnO^-, graphene oxide$^-$, and carbon nano-tubes-based sensors. BP and its single nuclear layer (phosphorene) have demonstrated an incredible chance to be viably utilized for gas identification like other two-dimensional materials like graphene, MoS_2, and so on [52, 53]. Cui *et al.* [54] announced up to 20 ppb level detection affectability in their mechanically peeled phosphorene nanosheets (PNS)-based FET gadget in a dry air climate.

Biosensing is another significant part of two-dimensional transition metal dichalcogenides that has proven critical development recently. Being bigger regarding their surface area, two-dimensional transition metal dichalcogenides gives immobilization of a lot of biomolecules per unit area and can bring about high proficiency biosensors prompting magnificent location of biomolecules including DNA, glucose, dopamine, and hydrogen peroxide, just to give some examples [55 - 57]. When contrasted with the three-dimensional mass materials and one-dimensional nanomaterials, for example, silicon nanowires or carbon nano-tubes-based biosensors, two-dimensional transition metal dichalcogenides offer high affectability and high gadget adaptability just as high adaptability and straightforwardness.

Energy Storage Applications of Polyoxometalate Modified Inorganic-organic Nanocomposite Materials

Polyoxometalates have risen as a promising material for the improvement of advanced gadgets. Their diverse scope of physical and chemical properties combined with fascinating redox performance make polyoxometalates engaging for an assortment of uses. The immediate utilization of these particles has been troublesome because of their high dissolvability and accordingly much work has been done on the readiness of polyoxometalate films immobilized on an assortment of substrates or assistance materials. For the manufacture of polyoxometalate electrodes, the utilization of exceptionally conductive natural substrates, for example, carbon or conducting organic polymers has been powerful. An assortment of hearty approaches exists for the creation of polyoxometalate altered natural cathodes: (i) The solid communication among polyoxometalates and carbon can be exploited to successfully unite polyoxometalate groups on a wide assortment of carbon allotropes through chemisorption. (ii) The oxidizing nature and acidic properties of hetero-polyacids give the ideal conditions to polymerizing monomers, for example, aniline, thiophene, and pyrrole to trap the polyoxometalate clusters in a COP network. (iii) The strong -ve charge of polyoxometalate anions permits them to be joined with emphatically charged polyelectrolytes for the congregation of electrostatically balanced label-by-label thin flicks on an assortment of natural substrates.

Label-by-label specifically gives off an impression of being the most important, taking into account that this technique is basic and flexible, yet considers the exact power over the structure and usefulness of resultant nanometer-scale multi-layer thin flicks. The quick reversible redox reactions of polyoxometalates combined with the physical capacitive impacts of carbon materials or the essential redox movement of COPs empowers the successful utilization of polyoxometalate composites in an assortment of energy stockpiling applications. These composite materials can be utilized to upgrade the discharge limit of lithium-ion batteries and increment the energy density of electrochemical capacitors. The creation of polyoxometalate hybrid materials for energy stockpiling is as yet a rising field and there are numerous open doors for additional exploration [58].

Graphene for Energy Storage and Conversion

It is assessed that the world should twofold its energy flexibly by 2050 [59], so it is of vital significance to grow new sorts of fuel sources. Contrasted with ordinary energy materials, carbon nanomaterials show unusual size and surface-subordinate (*e.g.,* morphological, electrical, optical, and mechanical) properties that upgrade energy-conversion execution. In particular, impressive endeavors have been consumed to abuse the special properties of graphene in elite energy-conversion gadgets, including sun-based cells and power devices. Alongside energy conversion, energy stockpiling is likewise of central significance, particularly for versatile applications. For some practical applications, high energy-stockpiling ability, high power conveyance capacity, and long life cycle are important. Attributable to the remarkable qualities of graphene, much exertion has been dedicated to the uses of graphene in elite supercapacitors and batteries.

Atomic Layer Deposition for Energy Storage

Atomic layer deposition was quickly created in the previous twenty years from a specialty innovation to a setup technique. In actuality, atomic layer deposition was at that point followed by the microelectronic business as the strategy of decision for the fabrication of next-generation metal-insulator-metal capacitors in DRAMs and to store gateway oxide in CMOS semiconductors [29]. Over the most recent couple of years, atomic layer deposition has additionally risen as the method of decision for the fabrication of different complex nanostructures. Atomic layer deposition previously arrived at the phase of development method is still in a quick turn of events. For application in energy stockpiling and transformation, a huge amount of complex nanostructured materials should be created in a sensible time.

The modern scale-up of the cycle is a significant test confronted these days by atomic layer deposition. To be sure, one of its highlights is that it is a moderate cycle and frequently homogenous covering of significant measure of nanomaterials is hard to be acquired. Indeed, not many milligrams of nanostructured materials can be generally handled. By and by, the atomic layer deposition community is now effectively examining the chance of scaling up as well as accelerating the cycles. For instance, the utilization of fluidized bed reactors [60, 61] in atomic layer deposition or the plan of novel reactors, for example, rotational reactors [62, 63] as of now demonstrated the capacity to measure multigram amounts of nanoparticles and nanotubes without breaking self-restricted surface reactions.

Polyaniline

Polyaniline is an ordinary electrode material for pseudo capacitors with high explicit capacitance and cycling dependability. It is likewise an exceptional cathode material for Li-ion batteries. Permeable carbon obtained from polyaniline carbonization and ensuing initiation measures has a high surface area and appropriate pore structure, with high nitrogen content, which can be utilized as an unrivalled carbon material in both energy stockpiling and conversion, especially to assist electrocatalysts in their function. Nevertheless, the wide use of polyaniline is more dependent on its exceptional formed-bonding structure just as the accessibility of plentiful N-active locales, which takes into consideration polyaniline to effectively couple with other terminal materials, such as carbonaceous materials, metal complexes or different polymers, bringing about composite materials with better execution over every part than synergistic impacts.

Polyaniline-based composite supercapacitors, charge-operated batteries, and electrocatalysts might have upgraded, or improved practices, in which polyaniline typically goes about as a permeable conductive help, defensive organization, or/and connective lattice on the outside of active electrode materials. Different methods have been utilized in setting up the composites with reasonable morphology, size, and structure, bringing about extraordinary advances in the previous many years. Polyaniline is a promising amazing terminal material for energy stockpiling and transformation gadgets [64].

CONCLUSION

Energy storage and conversion devices are very beneficial for all the purposes in daily life. This book chapter includes the different processes for energy conversion and storage like lithium-ion batteries, supercapacitors, fuel cells, polyoxometalates, polyaniline-based, two-dimensional dichalcogenides, chemical vapor deposition, thermal energy storage with phase change materials, etc. Some methods include the conversion of solar energy into chemical energy and further can be used for various purposes. Carbon nanotubes and graphene-based conducting polymers are also useful. Energy storage and conversion methods will be used in the future also and further modifications can also be needed as per future requirements and modifications.

CONSENT FOR PUBLICATION

Not applicable.

CONFLICT OF INTEREST

The authors declare no conflict of interest, financial or otherwise.

ACKNOWLEDGEMENTS

One of the authors Ms. Pinki wishes to express gratitude to the Council of Scientific and Industrial Research (CSIR), New Delhi, India, and University Grant Commission (UGC), New Delhi, India for financial assistance in the form of JRF vide letter no. 09/105(0308)/2020-EMR-I.

REFERENCES

[1]   Tarascon JM, Armand M. Issues and challenges facing rechargeable lithium batteries. Nature 2001; 414(6861): 359-67.
        [http://dx.doi.org/10.1038/35104644] [PMID: 11713543]

[2]   Armand M, Tarascon JM. Building better batteries. Nature 2008; 451(7179): 652-7.
        [http://dx.doi.org/10.1038/451652a] [PMID: 18256660]

[3]   Gregory DH. Lithium nitrides as sustainable energy materials. Chem Rec 2008; 8(4): 229-39.
        [http://dx.doi.org/10.1002/tcr.20151] [PMID: 18752308]

[4]   Song Z, Zhou H. Towards sustainable and versatile energy storage devices: an overview of organic electrode materials. Energy Environ Sci 2013; 6(8): 2280-301.
        [http://dx.doi.org/10.1039/c3ee40709h]

[5]   Armaroli N, Balzani V. The future of energy supply: Challenges and opportunities. Angew Chem Int Ed 2007; 46(1-2): 52-66.

[http://dx.doi.org/10.1002/anie.200602373] [PMID: 17103469]

[6]   VijayaVenkataRaman S, Iniyan S, Goic R. A review of climate change, mitigation and adaptation. Renew Sustain Energy Rev 2012; 16(1): 878-97.
[http://dx.doi.org/10.1016/j.rser.2011.09.009]

[7]   Alexiadis A. Global warming and human activity: A model for studying the potential instability of the carbon dioxide/temperature feedback mechanism. Ecol Modell 2007; 203(3-4): 243-56.
[http://dx.doi.org/10.1016/j.ecolmodel.2006.11.020]

[8]   Karakurt I, Aydin G, Aydiner K. Sources and mitigation of methane emissions by sectors: A critical review. Renew Energy 2012; 39(1): 40-8.
[http://dx.doi.org/10.1016/j.renene.2011.09.006]

[9]   Fernandes D, Pitié F, Cáceres G, Baeyens J. Thermal energy storage: "How previous findings determine current research priorities". Energy 2012; 39(1): 246-57.
[http://dx.doi.org/10.1016/j.energy.2012.01.024]

[10]   Dincer I, Rosen MA. Energetic, exergetic, environmental, and sustainability aspects of thermal energy storage for sustainable energy consumption. Netherlands: Springer 2007; pp. 23-46.

[11]   Mahatha SK, Patel KD, Menon KSR. Electronic structure investigation of MoS_2 and $MoSe_2$ using angle-resolved photoemission spectroscopy and *ab initio* band structure studies. J Phys Condens Matter 2012; 24(47): 475504.
[http://dx.doi.org/10.1088/0953-8984/24/47/475504] [PMID: 23110779]

[12]   Hwang WS, *et al.* Transistors with chemically synthesized layered semiconductor WS_2 exhibiting 10^5 room temperature modulation and ambipolar behavior. Appl Phys Lett 2012; 101: 013107.
[http://dx.doi.org/10.1063/1.4732522]

[13]   Vogt P, De Padova P, Quaresima C, *et al.* Silicene: compelling experimental evidence for graphenelike two-dimensional silicon. Phys Rev Lett 2012; 108(15): 155501.
[http://dx.doi.org/10.1103/PhysRevLett.108.155501] [PMID: 22587265]

[14]   Pakdel A, Zhi C, Bando Y, Golberg D. Low-dimensional boron nitride nanomaterials. Mater Today 2012; 15(6): 256-65.
[http://dx.doi.org/10.1016/S1369-7021(12)70116-5]

[15]   Ni Z, Liu Q, Tang K, *et al.* Tunable bandgap in silicene and germanene. Nano Lett 2012; 12(1): 113-8.
[http://dx.doi.org/10.1021/nl203065e] [PMID: 22050667]

[16]   Song L, Balicas L, Mowbray DJ, *et al.* Anomalous insulator-metal transition in boron nitride-graphene hybrid atomic layers. Phys Rev B Condens Matter Mater Phys 2012; 86(7): 075429.
[http://dx.doi.org/10.1103/PhysRevB.86.075429]

[17]   Ci L, Song L, Jin C, *et al.* Atomic layers of hybridized boron nitride and graphene domains. Nat Mater 2010; 9(5): 430-5.
[http://dx.doi.org/10.1038/nmat2711] [PMID: 20190771]

[18]   Liu Z, Song L, Zhao S, *et al.* Direct growth of graphene/hexagonal boron nitride stacked layers. Nano Lett 2011; 11(5): 2032-7.
[http://dx.doi.org/10.1021/nl200464j] [PMID: 21488689]

[19]   Geim AK, Novoselov KS. The rise of graphene. Nat Mater 2007; 6(3): 183-91.
[http://dx.doi.org/10.1038/nmat1849] [PMID: 17330084]

[20]   Allen MJ, Tung VC, Kaner RB. Honeycomb carbon: a review of graphene. Chem Rev 2010; 110(1): 132-45.
[http://dx.doi.org/10.1021/cr900070d] [PMID: 19610631]

[21]   Long DL, Tsunashima R, Cronin L. Polyoxometalates: building blocks for functional nanoscale systems. Angew Chem Int Ed 2010; 49(10): 1736-58.
[http://dx.doi.org/10.1002/anie.200902483] [PMID: 20131346]

[22]   Liu S, Tang Z. Polyoxometalate-based functional nanostructured films: Current progress and future prospects. Nano Today 2010; 5(4): 267-81.
[http://dx.doi.org/10.1016/j.nantod.2010.05.006]

[23]   Luo X, Li F, Xu B, Sun Z, Xu L. Enhanced photovoltaic response of the first polyoxometalate-modified zinc oxide photoanode for solar cell application. J Mater Chem 2012; 22(30): 15050-5.
[http://dx.doi.org/10.1039/c2jm16018h]

[24]   Sadakane M, Steckhan E. Electrochemical properties of polyoxometalates as electrocatalysts. Chem Rev 1998; 98(1): 219-38.
[http://dx.doi.org/10.1021/cr960403a] [PMID: 11851504]

[25]   Wang H, Hamanaka S, Nishimoto Y, *et al.* In operando X-ray absorption fine structure studies of polyoxometalate molecular cluster batteries: polyoxometalates as electron sponges. J Am Chem Soc 2012; 134(10): 4918-24.
[http://dx.doi.org/10.1021/ja2117206] [PMID: 22352694]

[26]   Gómez-Romero P, Cuentas-Gallegos K, Lira-Cantú M, Casañ-Pastor N. Hybrid nanocomposite materials for energy storage and conversion applications. J Mater Sci 2005; 40(6): 1423-8.
[http://dx.doi.org/10.1007/s10853-005-0578-y]

[27]   George SM. Atomic layer deposition: an overview. Chem Rev 2010; 110(1): 111-31.
[http://dx.doi.org/10.1021/cr900056b] [PMID: 19947596]

[28]   Knez M, Nielsch K, Niinistö L. Synthesis and Surface Engineering of Complex Nanostructures by Atomic Layer Deposition. Adv Mater 2007; 19(21): 3425-38.
[http://dx.doi.org/10.1002/adma.200700079]

[29]   Hwang CS. (2011). in Atomic Layer Deposition of Nanostructured Materials, (Eds: N. Pinna, M. Knez), Wiley-VCH 2011; p. 161.

[30]   Pinna N, Knez M. Atomic Layer Deposition of Nanostructured Materials. Wiley-VCH 2011.
[http://dx.doi.org/10.1002/9783527639915]

[31]   Wang K, Huang J, Wei Z. Conducting polyaniline nanowire arrays for high-performance supercapacitors. J Phys Chem C 2010; 114(17): 8062-7.
[http://dx.doi.org/10.1021/jp9113255]

[32]   Louarn G, Lapkowski M, Quillard S, Pron A, Buisson JP, Lefrant S. Vibrational properties of polyaniline e isotope effects. J Phys Chem 1996; 100(17): 6998-7006.
[http://dx.doi.org/10.1021/jp953387e]

[33]   Palacín MR. Recent advances in rechargeable battery materials: a chemist's perspective. Chem Soc Rev 2009; 38(9): 2565-75.
[http://dx.doi.org/10.1039/b820555h] [PMID: 19690737]

[34]   Recham N, Dupont L, Courty M, *et al.* Ionothermal Synthesis of Tailor-Made LiFePO$_4$ Powders for Li-Ion Battery Applications. Chem Mater 2009; 21(6): 1096-107.
[http://dx.doi.org/10.1021/cm803259x]

[35]   Russo P, *et al.* Effective fire extinguishing systems for lithium-ion battery. Chem Eng Trans 2018; 67: 727-32.

[36]   Lior N. The current status and possible sustainable paths to energy ''generation'' and Use. In nuclear & renewable energy conference (INREC), 2010 1st International. 2010.

[37]   Jeon J, *et al.* Application of PCM thermal energy storage system to reduce building energy consumption. J Therm Anal Calorim 2012; 1-10.

[38]   Mehling H, Cabeza L. Phase change materials and their basic properties Thermal energy storage for sustainable energy consumption. Netherlands: Springer. 2007; 234: pp. 257-77.

[39]   Ewing D. An investigation of the application of phase change materials in practical thermal

management systems. Clemenson University 2012.

[40]   Sharma SD, Sagara K. Latent heat storage materials and systems: a review. Int J Green Energy 2005; 2(1): 1-56.
[http://dx.doi.org/10.1081/GE-200051299]

[41]   Sabbah R, Kizilel R, Selman JR, Al-Hallaj S. Active (air-cooled) vs. passive (phase change material) thermal management of high power lithium-ion packs: Limitation of temperature rise and uniformity of temperature distribution. J Power Sources 2008; 182(2): 630-8.
[http://dx.doi.org/10.1016/j.jpowsour.2008.03.082]

[42]   Bailey E, Ray NMT, Hector AL, Crozier P, Petuskey WT, McMillan PF. Mechanical Properties of Titanium Nitride Nanocomposites Produced by Chemical Precursor Synthesis Followed by High-P,T Treatment. Materials (Basel) 2011; 4(10): 1747-62.
[http://dx.doi.org/10.3390/ma4101747] [PMID: 28824105]

[43]   Xu L, Li S, Zhang Y, Zhai Y. Synthesis, properties and applications of nanoscale nitrides, borides and carbides. Nanoscale 2012; 4(16): 4900-15.
[http://dx.doi.org/10.1039/c2nr30598d] [PMID: 22782140]

[44]   Zhang XQ, Lin CH, Tseng YW, Huang KH, Lee YH. Synthesis of lateral heterostructures of semiconducting atomic layers. Nano Lett 2015; 15(1): 410-5.
[http://dx.doi.org/10.1021/nl503744f] [PMID: 25494614]

[45]   Serna MI, Yoo SH, Moreno S, *et al.* Large-Area Deposition of MoS $_2$ by Pulsed Laser Deposition with *In Situ* Thickness Control. ACS Nano 2016; 10(6): 6054-61.
[http://dx.doi.org/10.1021/acsnano.6b01636] [PMID: 27219117]

[46]   Dan Y, Lu Y, Kybert NJ, Luo Z, Johnson ATC. Intrinsic response of graphene vapor sensors. Nano Lett 2009; 9(4): 1472-5.
[http://dx.doi.org/10.1021/nl8033637] [PMID: 19267449]

[47]   Lu G, Park S, Yu K, *et al.* Toward practical gas sensing with highly reduced graphene oxide: a new signal processing method to circumvent run-to-run and device-to-device variations. ACS Nano 2011; 5(2): 1154-64.
[http://dx.doi.org/10.1021/nn102803q] [PMID: 21204575]

[48]   Late DJ, Huang YK, Liu B, *et al.* Sensing behavior of atomically thin-layered MoS₂ transistors. ACS Nano 2013; 7(6): 4879-91.
[http://dx.doi.org/10.1021/nn400026u] [PMID: 23713986]

[49]   Kalantar-zadeh K, Ou JZ. Biosensors Based on Two-Dimensional MoS $_2$. ACS Sens 2016; 1(1): 5-16.
[http://dx.doi.org/10.1021/acssensors.5b00142]

[50]   Li H, Yin Z, He Q, *et al.* Fabrication of single- and multilayer MoS₂ film-based field-effect transistors for sensing NO at room temperature. Small 2012; 8(1): 63-7.
[http://dx.doi.org/10.1002/smll.201101016] [PMID: 22012880]

[51]   Donarelli M, Prezioso S, Perrozzi F, *et al.* Response to NO₂ and other gases of resistive chemically exfoliated MoS₂-based gas sensors. Sens Actuators B Chem 2015; 207: 602-13.
[http://dx.doi.org/10.1016/j.snb.2014.10.099]

[52]   Xia F, Wang H, Jia Y. Rediscovering black phosphorus as an anisotropic layered material for optoelectronics and electronics. Nat Commun 2014; 5(1): 4458.
[http://dx.doi.org/10.1038/ncomms5458] [PMID: 25041752]

[53]   Kou L, Frauenheim T, Chen C. Phosphorene as a Superior Gas Sensor: Selective Adsorption and Distinct $I-V$ Response. J Phys Chem Lett 2014; 5(15): 2675-81.
[http://dx.doi.org/10.1021/jz501188k] [PMID: 26277962]

[54]   Cui S, Pu H, Wells SA, *et al.* Ultrahigh sensitivity and layer-dependent sensing performance of phosphorene-based gas sensors. Nat Commun 2015; 6(1): 8632.
[http://dx.doi.org/10.1038/ncomms9632] [PMID: 26486604]

[55]   Wang X, Nan F, Zhao J, Yang T, Ge T, Jiao K. A label-free ultrasensitive electrochemical DNA sensor based on thin-layer MoS_2 nanosheets with high electrochemical activity. Biosens Bioelectron 2015; 64: 386-91.
[http://dx.doi.org/10.1016/j.bios.2014.09.030] [PMID: 25262063]

[56]   Mao K, Wu Z, Chen Y, Zhou X, Shen A, Hu J. A novel biosensor based on single-layer MoS2 nanosheets for detection of Ag+. Talanta 2015; 132: 658-63.
[http://dx.doi.org/10.1016/j.talanta.2014.10.026] [PMID: 25476360]

[57]   Huang J, He Y, Jin J, Li Y, Dong Z, Li R. A novel glucose sensor based on MoS_2 nanosheet functionalized with Ni nanoparticles. Electrochim Acta 2014; 136: 41-6.
[http://dx.doi.org/10.1016/j.electacta.2014.05.070]

[58]   Genovese M, Lian K. Polyoxometalate modified inorganic–organic nanocomposite materials for energy storage applications: A review. Curr Opin Solid State Mater Sci 2015; 19(2): 126-37.
[http://dx.doi.org/10.1016/j.cossms.2014.12.002]

[59]   Doung TG. "FY 2002 Annual Progress Report for Energy Storage Research and Development" (Freedom CAR and Vehicle Technologies Program, US Department of Energy, Washington, DC, 2003). 2003.

[60]   King DM, *et al.* Atomic layer deposition on particles using a fluidized bed reactor with in situ mass spectrometry. Surf Coat Technol 201(22–23):9163–9171.

[61]   Liang X, King DM, Li P, Weimer AW. Low-Temperature Atomic Layer-Deposited TiO $_2$ Films with Low Photoactivity. J Am Ceram Soc 2009; 92(3): 649-54.
[http://dx.doi.org/10.1111/j.1551-2916.2009.02940.x]

[62]   McCormick JA, Rice KP, Paul DF, Weimer AW, George SM. Analysis of Al_2O_3 Atomic Layer Deposition on ZrO_2 Nanoparticles in a Rotary Reactor. Chem Vap Depos 2007; 13(9): 491-8.
[http://dx.doi.org/10.1002/cvde.200606563]

[63]   Wilson CA, McCormick JA, Cavanagh AS, Goldstein DN, Weimer AW, George SM. Tungsten atomic layer deposition on polymers. Thin Solid Films 2008; 516(18): 6175-85.
[http://dx.doi.org/10.1016/j.tsf.2007.11.086]

[64]   Wang H, Lin J, Shen ZX. Polyaniline (PANi) based electrode materials for energy storage and conversion. J Sci Adv Mater Devices 2016; 1(3): 225-55.
[http://dx.doi.org/10.1016/j.jsamd.2016.08.001]

CHAPTER 10

Investigation on Various Polymer Electrolytes for Development of Dye Sensitized Solar Cell

Priyanka Chawla[1,*], **Shivangi Trivedi**[1] and **Kumari Pooja**[1]

[1] *Department of Chemistry, CMP Degree College, University of Allahabad, Uttar Pradesh 211002, India*

Abstract: Dye sensitized solar cells (DSSCs) based on TiO2 nanoparticles film have attracted extensive attention from both industry and academia. Generally, the liquid electrolyte is used in dye sensitized solar cells, but the vaporization of liquid electrolyte hinders its commercialization as its affects its stability. And also the reduction in performance of dye sensitized solar cells was observed due to electron recombination in semiconductor liquid electrolyte interfaces. The situation worsens when the photoanode is in contact with the vaporization of electrolyte solution that affects the charge distribution at the semi conductor electrolyte interface and initiates photo corrosion on the photoanode. With the finding of ionic conductivity in polymer, electrolytes complexed with salt give a breakthrough to the development of DSSC devices. Various types of electrolytes have been developed and tested in different DSSCs configurations to overcome this problem. Among all polymer electrolytes, PEO (Polyethylene oxide) based polymer electrolyte has shown excellent performance in different electrochemical application areas. In DSSCs, it is also considered a novel candidate due to its excellent ability to form complexes with ionic salts. Poly(vinyl alcohol) (PVA) is also a promising candidate acting as a host polymer due to its inherent characteristics like high mechanical strength, good tensile strength, high temperature resistance, non toxicity, good optical properties and high hydrophilicity. PVA have a large extent of poly hydroxyl group, which makes PVA highly hydrophile. It also offers other advantages like excellent chemical stability, ease of preparation, and flexibility. In the present paper, we review different types of polymer electrolytes which have been used for improving the performance and stability of DSSCs.

Keywords: Dye Sensitized Solar Cell, Polymer Electrolyte, Polyethylene Oxide, Polyvinyl Alcohol, Chitosan.

INTRODUCTION

Solar cells are the devices that convert solar energy into electricity. The conventional solar cell device is silicon based solar cells with high conversion eff-

* **Corresponding author Priyanka Chawla:** Department of Chemistry, CMP Degree College, University of Allahabad, Uttar Pradesh 211002, India; E-mail: ccmpau@gmail.com

Arti Srivastava, Mridula Tripathi, Kalpana Awasthi and Subhash Banerjee (Eds.)

iciency. But the high cost of developing these silicon-based solar cells limits their widespread application. Therefore in last few years, new concepts of solar cells have been considered. These technologies mainly focussed on concentrating photovoltaic technology, organic solar cells, dye sensitized solar cells and novel emerging solar cell concepts, and together these are known as third generation solar cells. Because of their low cost materials and easy fabrication, these technologies are expected to take a significant share in the fast growing photovoltaic area.

Since the breakthrough work by Gratzel in 1991, Dye sensitized solar cells (DSSCs) have attracted the interest of both industry and academia for future clean energy. The four major components of DSSCs are sensitizers, nanocrystalline porous semiconductor based photoanode, electrolytes and counter electrode [1 - 3]. The performance of a DSSC is affected by the properties of metal oxide, as well as the choice of electrolyte and dye. Therefore the researchers working in the field of DSSCs are focusing their attention on improving one or the other component of DSSCs so that efficiency comparable to silicon based solar cell can be achieved.

The liquid electrolyte, which is the most important component of DSSCs, hinderers its stability as the electrolyte leaks and vaporizes with time. Thus scientists are focusing their attention on the solidification of an electrolyte such as inorganic or organic hole conductors, ionic liquids, polymer electrolytes [4 - 6]. Polymer electrolyte is the most important ionic conductor that is used in electrochemical devices. Polymer electrolytes are solid ionic conductors that are prepared by the dissolution of salts in a suitable high molar mass polymer containing polyether units. For obtaining good performance from electrochemical devices, the polymer electrolyte used must have excellent properties such as good ionic conductivity, and thermal, mechanical and electrochemical stabilities. Among all these properties, we have focused our attention on enhancing ionic conductivity of the polymer electrolyte. The ionic conductivity of a polymer electrolyte is given by the equation:

$$\sigma = \eta \mu e$$

Where η is number density μ is the mobility of charge carriers, and e is the elementary charge. From the equation, it can be seen that η and μ are two important parameters that control the ionic conductivity of polymer electrolytes. Thus, it is essential that η and μ be determined quantitatively.

Polymeric electrolytes are materials of special interest because they show many advantageous properties when used in devices such as low cost, easy film formation *etc.* The studies on polymer electrolytes are relatively motivating to physicists, chemists and engineers for their fundamental physical properties and potential applications in many electrochemical devices such as batteries, fuel cells, super capacitors, sensors and display devices. They are in crystalline and amorphous phases. The existence of an amorphous phase and lower values of glass transition temperature are accountable for ion conduction in such systems. These materials have various advantages over liquid electrolytes, such as; corrosion, self-discharge, bulky design, miniaturization *etc.* Various approaches have been adopted for the synthesis of new polymer electrolytes exhibiting higher ionic conductivity at ambient temperature such as polymer blends, copolymers, comb branch polymer, cross-linked networks, addition of plasticizers, addition of ceramic filler, and use of a larger anion of dopant salt (acid) [7 - 10]. Though the highest reported efficient DSSC contains a volatile organic solvent which suffers from the major drawback of leakage and vaporization of organic solvent which hinders its long-term practical operation. Moreover, the corrosion of iodine on platinum electrolytes is also an additional barrier. Therefore, by using polymer electrolytes, one can overcome these problems [11 - 14].

BASIC PRINCIPLE OF DYE-SENSITIZED SOLAR CELLS

The disparity between DSSCs and p-n junction solar cells is about their components and working. In p-n junction solar cells, the semiconductor performs both the tasks of light harvesting and charge carrier transport, while in DSSCs these two functions are performed separately. Moreover, the solar energy conversion mechanisms in DSSCs are the interfacial processes while in p-n junction cells these bulk processes. Hence most studies on DSSC are made to understand the prevailing role of electron transfer dynamics and kinetics at nanocrystalline metal oxide/sensitizer/electrolyte interfaces. Though researchers are performing studies on this subject, still not much understanding of the kinetics of the interfacial processes has been made. If we properly understand the kinetics of interfacial processes, we can improve the efficiency of DSSC and scale up their manufacturing.

Dye sensitized solar cells (DSSCs) are nanostructured photoelectrochemical device in which photons are absorbed by the sensitizers attached to the large band gap semiconductor oxide. The conversion of photonic energy into electricity takes place by the transfer of electrons from the excited dye molecule to the conduction band of the semiconductor oxide. The electron moves from the semiconductor oxide to the current collector and the external circuit. In the pores, there is a redox mediator which ensures that the oxidized dye species are continuously regenerated over and over again and the cycle is not stopped as shown in Fig. (1).

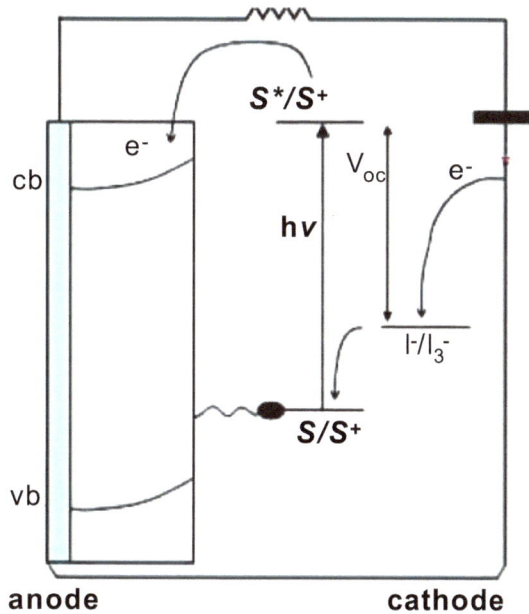

Fig. (1). Energy diagram of DSSC.

PREPARATION OF POLYMER ELECTROLYTE FILM

Following methods have been adopted by researchers to prepare polymer electrolyte films

• Solution Cast technique

• Sol-Gel technique in a liquid electrolyte medium

• Polymerization in the presence of liquid electrolyte medium

• Phase inversion method

• Soaking of liquid electrolyte in polymer matrix

Solution Cast Technique

For the preparation of polymer electrolyte films, various methods are adopted but the solution cast technique is the most common, easy and cheapest. In this technique, stoichiometric proportions of the polymer and additives such as salt or acid are dissolved in a suitable solvent. The two solutions are mixed and stirred together vigoursly for a particular duration of time at an elevated temperature to form a homogeneous viscous solution. After that filler particles are dispersed heterogeneously to achieve the composite polymer electrolyte and continuously stirred for 8-10 hours. Then this solution is cast on a Teflon or polycarbonate or glass moulds to remove the solvent by a slow evaporation process in the air at room temperature. These solution cast films are dried at a constant 30°C in a BOD incubator before vacuum drying to obtain solvent free polymer electrolyte films.

The morphology of the prepared polymer electrolytes is significantly affected by the nature of the solvent, the rate of removal of the solvent and traces of residual solvent. And other factors such as impurities, atmospheric conditions and thermal treatment put an adverse impact over the morphology of the prepared polymer electrolytes. Therefore precautions are very important while performing this technique for the preparation of polymer complexes. However, in practice, there is no universally acceptable common criterion. Various workers also adopted several other techniques but they could not become popular due to their drawbacks. If proper control is taken of the entire process and the appropriate solvent is taken then it is possible to obtain polymer electrolyte films of desired thickness with consistent reproducibility. Given the above, the solution cast technique has been used in the present investigation (Fig. **2**).

PEO BASED POLYMER ELECTROLYTE FOR DSSCs

PEO (Polyethylene oxide) based polymer electrolyte has shown excellent performance in the different electrochemical application areas. In DSSCs, it is also consider a novel candidate due to their excellent ability to from complexes with the ionic salts. PEO shows good salvation properties due to the presence of unpaired electrons on the ether oxygen atoms which lead to good ionic ability [15, 16].

The first DSSC based on PEO solid electrolyte without any additives and dopants was reported in the year 1999 by Nogueira *et al.* in which they used PEO-epychlomer with sodium iodide (NaI) and iodine (I$_2$) as a redox couple and conducting poly(o- methoxy aniline) as a sensitizer. The reported efficiency of their cell was 1.6x 10^{-4}% at 120 mW cm^{-2} such a low efficiency can be attributed to a high crystalline matrix of polymer which is responsible for low ionic conductivity of polymer electrolyte in comparison with liquid electrolyte [17].

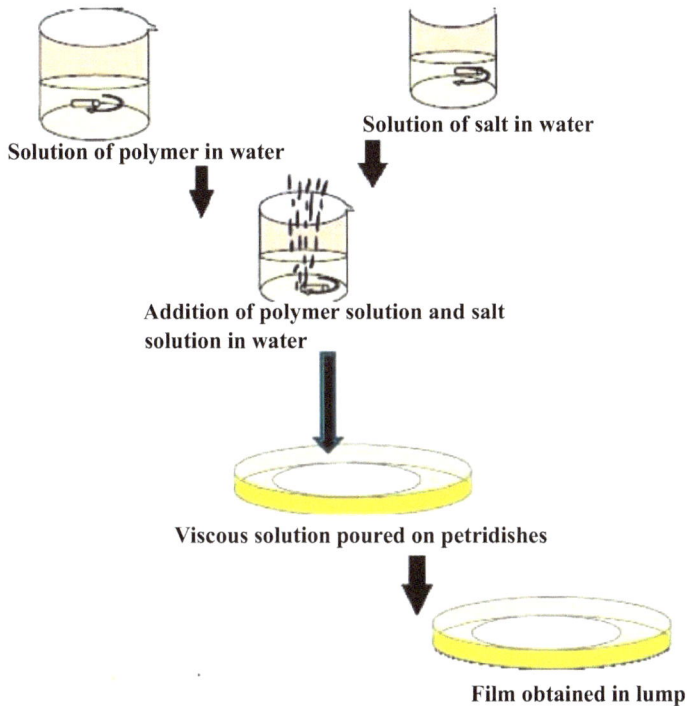

Fig. (2). Synthesis of nano composite polymer electrolyte films by Solution Cast Techniq.

In the year 2005, Kim and co -workers reported a DSSC containing PEO/NaI:I$_2$ polymer electrolyte with ruthenium based dye which shows an efficiency of 0.07% at 10 mW cm^{-2} light intensity. Another group reported an efficient DSSC based on PEO: KI/I$_2$ system with 2.04% efficiency at 100 mW cm^{-2} light intensity. As we can see from the proposed DSSCs that overall efficiency has already reached the limit for the system based on the polymer iodide salt/ iodine complex. The use of only PEO as the electrolyte could not show higher values than this. For further improvement in the efficiency of the DSSC it is necessary to modify the electrolyte by adding various additives such as plasticizers, copolymers, ionic liquids and inorganic nanoparticles [18].

To improve the conductivity of PEO polymer electrolyte based DSSC, reduction of crystallinity of PEO or reducing its glass transition temperature is very important. Generally, plasticizers of low molecular weight are incorporated in small amounts into a polymeric matrix to increase its segmental motion which leads to a degree of disorder in the polymer matrix which is necessary for further improvement in electrical conductivity and its device application. Many popular plasticizers like polyethylene glycol (PEG), polypropylene glycol (PPG), ethylene carbonate (EC), propylene carbonate (PC) *etc.* have already shown their importance in enhancing ionic conductivity by reducing crystallinity. Ileperuma *et al.* fabricated a DSSC using EC, PC plasticizers doped PEO: KI: I_2 polymer electrolyte system with high conductivity 2.2 x 10^{-3} S/Cm and DSSC efficiency 0.6% at 100 mW/cm² [19]. Freitas *et al.* reported an efficient and stable DSSC with an efficiency of 3% using plasticizers butyrolactone (BL) in PEC containing LiI and I_2 system [20]. An efficient DSSC was claimed by Flores and co- workers by applying plasticizer poly (ethyleneglycol) dibenzoate (PEG-diB) in PEO epychlomer (PEC): NaI: I_2 matrix. They obtained an average efficiency of 0.9% per cell operated in outdoor conditions. The overall increase in the efficiency of the PEO based polymer electrolyte is due to a reduction in crystallinity of the polymer matrix by adding plasticizers [21].

The addition of different polymers in the PEO matrix is also a well-known method due to its lower glass transition temperature and improved amphoursity. Prabakaran *et al.* prepared a polymer blend electrolyte membrane for DSSCs based on PEO and poly (vinylidene fluoride- co- hexaflouro propylene) (PVDF-HFP). Along with that, nanosized titanium oxide fillers were also added to the membrane to improve its conductivity. The ionic conductivity of around 7.21 X 10^{-4} S/cm was obtained for the film. The photovoltaic characteristics of the fabricated cell showed an open circuit voltage 0.71 V and the best efficiency was about 2.84% [22].

Rao *et al.* prepared a sequence of polymer electrolytes by the addition of different weight percentages of acetamide in PEO/PEGDME (Poly (ethylene glycol) dimethyl ether) blended polymer matrix and NaI/I_2 redox couple. The maximum conversion efficiency of 5.29% was achieved for PEO/PEGDME with 10 wt% acetamide under 100 mW/cm² illumination and ionic conductivity in the range of 10^{-3} S/cm. The improved performance of the cell is due to an increase in ionic conductivity of the polymer electrolyte which is attributed due to the addition of acetamide in the blended PEO polymer matrix. The addition of acetamide increases the amorphous nature of the polymer and lowers its glass transition temperature [23].

The addition of nanosized inorganic fillers is also one of the successful approaches to enhance the ionic conductivity and mechanical properties of polymer electrolyte. Addition of nanosized inorganic fillers is considered better in comparison to the addition of only additives as these additives may deteriorate the mechanical properties of the polymer electrolyte but these nanofillers not only improve the mechanical properties but also enhance the conductivity and relaxation properties of the electrolyte by affecting the mobility of polymer chain segments and chain packing by favouring dissociation of salt to form free ions that contribute in conduction process. Such formed polymer electrolytes consisting of nanoparticles are called polymer nanocomposite electrolytes. Various nanofillers such as TiO_2, SiO_2, Al_2O_3, *etc.* are incorporated in the electrolyte film for improved conductivity [24, 25].

PVA BASED POLYMER ELECTROLYTE FOR DSSCs

Poly(vinyl alcohol) (PVA) is also a promising candidate acting as a host polymer due to its inherent characteristics like high mechanical strength, good tensile strength, high temperature resistance, nontoxicity, good optical properties and high hydrophilicity. PVA has a large extent of poly hydroxyl group which makes PVA high hydrophile. It offers other advantages like excellent chemical stability, ease in preparation, and good flexibility. It is also used for the development of fuel cell and electrical double- layer capacitor due to the high storage capacity [26, 27].

Aziz *et al.* prepared PVA based gel polymer electrolyte with iodide slats (KT and Bu4NI) for fabrication of DSSC. The X –ray diifraction technique confirms the amorphous nature of the prepared film. Impedance spectroscopic technique was employed to determine the conductivity of the prepared gel polymer electrolyte. The electrolyte with 30% KI and 70% Bu4NI showed a conductivity of about 5.25 X 10 $^-$3mS/cm. The DSSC with this electrolyte showed an efficiency of about 5.8%. Synthetic dye was used for the fabrication of the cell system [28].

Senthil and group fabricated DSSC with $PVA/KI/I_2$ polymer electrolyte with hematite iron oxide nanoparticles (α-Fe_2O_3 NPs). They prepared pure and different weight (%) ratios (2,3,4 and 5% w.r.t PVA) of α-Fe_2O_3 NPs incorporated $PVA/KI/I_2$ polymer electrolyte film by solution cast method using DMSO as a solvent. AC impedance studies showed a significant increase in conductivity for α-Fe_2O_3 NPs incorporated polymer electrolyte than compared to pure $PVA/KI/I_2$. This is due to the reduction in the crystalline nature of the polymer which enhances the mobility of I^-/I_3^- redox couple. The highest conductivity of 1.167 X 10^{-4}S/ Cm was observed for 4% weight of α-Fe_2O_3 NPs incorporated $PVA/KI/I_2$ polymer electrolyte. The DSSC fabricated with this electrolyte showed an enhanced power conversion efficiency of 3.62% [29].

Hassan and co-workers investigated the performance of chlorophyll sensitized solar cell with gel electrolyte based on PVA with single iodide salt and potassium iodide (KI) and double salt (KI) and tetrapropyl ammonium iodide (TPAI). From the studies, it has been found that the double salt electrolyte system with PVA showed an efficiency of about 2% with Jsc 5.96 mA/cm^{-2}, Voc 0.58 V and FF 64% whereas the electrolyte system with only one salt showed the efficiency of about 1.77% [30].

Seni *et al.* investigated the affect of the addition of polymer materials such as PEG 1000, 4000 weight and PVA 60000 for fabricating a gel electrolyte. The results showed that the DSSC fabricated using PVA 60000 based gel electrolyte showed better performance in comparision to PEG 1000, 4000 based electrolyte [31]. Chawla and Co-Workers prepared a PVA based solid nanocomposite polymer electrolyte film for anthocyanin dye based DSSC. PVA based polymer electrolyte film was prepared using graphite filler and LiI and I_2 as a redox couple. The obtained conversion efficiency was about 1.65 under one sun condition [32].

To further improve the conductivity of polymer electrolytes, various approaches have been adopted such as polymer blending, the addition of inorganic/organic fillers, and the incorporation of dopants and plasticizers.

Katsaros *et al.* prepared solvent free composite polymer electrolyte consisting of high molecular mass polyethylene oxide (PEO) filled with titanium oxide as inorganic filler and containing $LiI:I_2$ salts. From the studies, it has been found that the introduction of TiO_2 in the polymer matrix produces dramatic morphological changes to the host polymer structures. From the atomic force microscope (AFM) pictures, it has been seen that with the introduction of TiO_2, the polymer chains are separated craeting voids into which the iodide and triiodide anions can easily migrate which overall increases the conductivity of the electrolyte film. All solid state dye sensitized solar cells fabricated using this electrolyte present high efficiencies (typical maximum incident to photon to current efficiency (IPCE) as high as 40% at 520 nm and overall conversion efficiency of about 0.96% (Voc=0.67 V, Jsc=2.05 mA/cm^2, FF=39%) under direct solar radiations [33].

Polymer blending is also one of the feasible techniques adopted for improving the performance of polymer electrolytes. Polymer blending is a process in which at least two polymers are mixed with or without any chemical bonding. A polymer blend is therfore a physical mixture of two or more polymers. A PMMA/PVDF blend based polymer electrolyte was used for lithiumion batteries. Buraidah *et al.* prepared a polymer blend electrolyte by mixing chitosan and poly(ethylene oxide) in different weight ratios. The X-ray diffraction technique showed the amophous nature of the prepared polymer electrolyte film. The impedance spectroscopy showed the electrical conductivity of the film in the range of 10^{-6} S/Cm. The DSSC prepared using the polymer electrolyte system showed an efficiency of about 0.78% under one sun condition [34].

Senthil and co-workers investigated the influence of guanine as an organic dopant in DSSC based on poly(vinyldinefluoride-co-hexafluropropylene) (PVDF-HFP)/ poly(ethylene oxide) (PEO) polymer blend electrolyte along with iodide salts potassium iodide (KI) and tetrabutylammonium iodide (TBAI) and iodine (I_2).

The film was prepared using the solution cast technique with DMF as a solvent. The PVDF-HFP/KI + TBAI/I_2 electrolyte showed an ionic conductivity value of 9.99×10^{-5} S/Cm. When PEO was blended with PVDF-HFP/KI + TBAI/I_2 the electrolyte showed ionic conductivity of about 4.53×10^{-5} S/Cm. However, a maximum ionic conductivity value of about 3.67×10^{-4} S/Cm was obtained when guanine was introduced into the above electrolyte system. This shows that guanine is an effective dopant to enhance the conductivity of polymer electrolyte system the obtained conversion efficiency was about 2.46%, respectively [35].

PVDF BASED POLYMER ELECTROLYTE FOR DSSCs

PVDF (Polyvinylidene fluoride) can be considered suitable for forming polymer electrolytes as it has some exceptional properties like thermal stability, chemical resistance, excellent mechanical strength and photo-electrochemical stability under potential application [36, 37].

Noor *et al.* fabricated dye-sensitized solar cells with a polymer electrolyte system composed of PVDF-HFP, potassium iodide, and equal amounts of ethylene carbonate and propylene carbonate. The electrolyte exhibits the highest ionic conductivity of 10^{-3} S/Cm. The DSSC was fabricated with the electrolyte film sandwiched between a TiO_2/dye photoelectrode and platinum-based counter electrode. Ruthenizer 535 (N3) dye was used as a sensitizer. The obtained solar efficiency was 2.2% with a fill factor 35% and current density of 8.16 mA cm^{-2} and an open circuit voltage of 0.76 V [38].

Perera and Vidanapathirana used gel polymer electrolyte with PVdF, ethylene carbonate, propylene carbonate, tetrahexylammonium iodide and iodine for DSSC application. The conductivity was found to be in the range of 10^{-3} S/Cm. The open circuit voltage and short circuit current density were 799 mV and 1.06 mA, respectively. The calculated fill factor was 0.62 and the efficiency was 2.08% [39].

Kannadhasan used pure PVDF-HFP/LiI/I_2 and 4-nitroaniline doped PVDF-HFP/LiI/I_2 based polymer electrolyte film for fabricating DSSC. Impedance studies revealed that 4-nitroaniline doped polymer electrolyte showed higher ionic conductivity (3.18×10^{-6} S/cm) compared to pure (1.88×10^{-7} S/cm) polymer electrolytes. This can be attributed to the better conduction of ions in the doped polymer electrolyte in comparison to that without doping. The conversion efficiency of pure and 4- nitoaniline doped polymer electrolyte-based DSSCs were 1.3% and 1.5%, respectively [40].

Sun *et al.* demonstrated the design and operation of DSSC based on multi-walled carbon nanotube counter electrode and a polymer electrolyte consisting of PVDF. Morphological studies have shown that the polymer electrolyte was found to be very highly porous which provided an optimized interfacial contact with the multi-walled carbon nanotube based counter-electrode. The efficiency of the cell was found to be 6.4% under one sun. Whereas when liquid electrolyte was used the cell efficiency was found to reduce gradually due to internal resistance caused by the reaction between liquid electrolyte and iodide and triodide [41].

CHITOSAN BASED POLYMER ELECTROLYTE FOR DSSCs

Chitosan is an important polymer and can be used as a polymer electrolyte for DSSCs. Chitosan has good film or membrane forming ability. This is an advantage as most of the polymer electrolytes are prepared in film form *via* the solution cast technique. For forming a polymer electrolyte film, the cation of the salt must co-ordinate electrostatically with the electron donor atom of the polymer [42]. The chitosan-based electrolyte films are formed to be homogenous and show high mechanical strength. Chitosan is biodegradable, biocompatible, odourless and non toxic. Chitosan is a linear polysaccharide and has the ability to form thin films, which is resistant to chemical attack, easy to cross link and has electrolytic properties [43]. Chitosan films are good liquid absorbers and hence can achieve good ionic conductivity when immersed in liquid electrolytes. This property is very important for energy storage devices. All these properties make chitosan a suitable candidate for forming liquid electrolyte films and can be used in DSSCs [44].

Aziz *et al.* have prepared proton conducting electrolytes based on phthaloylated chitosan with NH_4SCN as the doping salt. The highest room temperature conductivity was found to be 2.42 X 10^{-3}S/cm for the 30% weight salt containing sample [45]. Likewise Muthumeenal and co-workers measured the conductivity of polyethersulfon/ phthaloyl chitosan blended electrolyte dissolved in concentrated sulphuric acid. The room temperature conductivity was found to be 92 X 10^{-2} S/Cm. The high conductivity can be attributed due to the presence of sulfonic acid groups and the presence of polar groups in the pthaloyl chitosan [46].

Buraidah *et al.* fabricated DSSC using a chitosan-based polymer electrolyte and TiO_2 photoanode. The dye used was anthocyanin obtained by solvent extraction from Black rice and betalain from the callus of *Celosia pulmosa*. The chitosan 22.5 wt% NH_4I 50% showed the electrical conductivity of 3.43 x 10^{-5} S/Cm respectively at room temperature. It is observed that the circuit current density increases the conductivity of the electrolyte [47]. SNF Yusuf and co-workers prepared a phthaloylchitosan based gel polymer electrolyte with tetrapropy-lammonium iodide, Pr_4NI, as the salt and used it for DSSCs. The electrolyte with the composition of 15.7 wt% phthaloylchitosan, 31.7 wt% ethylene carbonate (EC) and 3.17 wt% propylene carbonate (PC), 19.0 wt% of Pr_4NI and 1.9 wt% iodine exhibits the highest room temperature ionic conductivity of about 5.27 X 10^{-3} S/Cm. The synthetic dye based DSSC fabricated with this electrolyte exhibits an efficiency of about 3.5% with Jsc of 7.38 mA/cm^2, Voc of 0.12 V and fill factor of 0.66 [48].

Maddu *et al.* fabricated DSSC using Cadmium sulphide (Cds) films asphotoanode and polymers blend used based on chitosan and polyethylene glycol (PEG) as a solid electrolyte. The Cds films were deposited on the ITO glass substrates by the chemical bath deposition method. The solid-state electrolyte was prepared by blending chitosan and PEG as a matrix, for the redox couple potassium iodide/iodine (KI/I_2) was used. It was found that the addition of KI/I_2 affected the conductivity of the electrolyte. The individual solar cells were formed into the sandwich structure ITO/Cds/Gel electrolyte/ITO glass [49, 50].

From the above, we can see that replacing the liquid electrolyte with a solid polymer electrolyte is one best way to improve the stability of dye-sensitized solar cells. Some of the best results of all the polymers taken into consideration are shown in Table **1**.

Table 1. Some of the best results of all the polymers taken in consideration.

DSSC Cell Structure	Conductivity	Efficiency	References
PEO: KI/I_2/Synthetic dye	10-3 s/cm	2.04	18
PEC: LiI/I2/Synthetic dye/ butyrolactone (BL) as plasticizer	10-3 s/cm	3	21
PEO/PEGDME/ NaI/I_2/synthetic dye	10-3 s/cm	5.29	23
PVA/KI/ Bu_4NI/synthetic dye	5.25 X 10^{-3}mS/cm	5.8	28
α-Fe_2O_3 NPs incorporated PVA/KI/I_2/Synthetic dye	1.167 X 10^{-4}S/ Cm	3.62	29
PVA/LiI/I_2/Natural dye anthocyanin/EC and PC as plasticizers	10^{-4}	1.65	33
PEO-PVDF-HFP/KI+ TBAI/I_2/Synthetic dye	3.67X 10^{-4} S/Cm	2.46	35
4-nitroaniline doped PVDF-HFP/LiI/I_2	3.18X 10^{-6}S/cm	1.5	40
Phthaloylchitosan based gel polymer electrolyte with tetrapropylammonium iodide, Pr_4NI,	5.27 X 10^{-3} S/Cm.	3.5	48

CONCLUSION

In order to improve the stability of DSSCs, liquid electrolytes are being replaced by solid polymer electrolytes. Polymer electrolytes offer several advantages over the conventional liquid electrolyte. Polymer electrolyte possesses transport properties that are comparable with some liquid ionic solutions, and also have some other interesting properties such as transparency, solvent free, light weight, flexibility, thin film forming stability, high ionic conductivity, easy processability, and wide electrochemical windows. The use of PEs enhances safety as they prevent some problems such as electrolyte leakage, internal shorting, use of corrosive solvent, production of harmful gases and the presence of non - combustible reaction products on the electrode surface. In this respect, polymer electrolytes are the most commonly used ionic conductors as they offer the advantages of solid state electrochemistry with the ease of processing. It may conclude that the polymer electrolyte helped in attaining the stability of the prepared DSSCs.

CONSENT FOR PUBLICATION

Not applicable.

CONFLICT OF INTEREST

The authors declare no conflict of interest, financial or otherwise.

ACKNOWLEDGEMENT

One of the authors, Dr. Priyanka Chawla is thankful to CSIR-HRDG India for the financial support.

REFERENCES

[1]   Hagfeldt A, Grätzel M. Molecular Photovoltaics. Acc Chem Res 2000; 33(5): 269-77.
      [http://dx.doi.org/10.1021/ar980112j] [PMID: 10813871]

[2]   Grätzel M. Photoelectrochemical cells. Nature 2001; 414(6861): 338-44.
      [http://dx.doi.org/10.1038/35104607] [PMID: 11713540]

[3]   Grätzel M. Solar energy conversion by dye-sensitized photovoltaic cells. Inorg Chem 2005; 44(20): 6841-51.
      [http://dx.doi.org/10.1021/ic0508371] [PMID: 16180840]

[4]   Stergiopoulos T, Arabatzis IM, Katsaros G, Falaras P. Binary Polyethylene Oxide/Titania Solid-State Redox Electrolyte for Highly Efficient Nanocrystalline TiO_2 Photoelectrochemical Cells. Nano Lett 2002; 2(11): 1259-61.
      [http://dx.doi.org/10.1021/nl025798u]

[5]   Katsaros G, Stergiopoulos T, Arabatzis IM, Papadokostaki KG, Falaras P. A solvent-free composite polymer/inorganic oxide electrolyte for high efficiency solid-state dye-sensitized solar cells. J Photochem Photobiol Chem 2002; 149(1-3): 191-8.
      [http://dx.doi.org/10.1016/S1010-6030(02)00027-8]

[6]   Kalaignan G, Kang M, Kang Y. Effects of compositions on properties of PEO–KI–I2 salts polymer electrolytes for DSSC. Solid State Ion 2006; 177(11-12): 1091-7.
      [http://dx.doi.org/10.1016/j.ssi.2006.03.013]

[7]   Armand M. Polymers with Ionic Conductivity. Adv Mater 1990; 2(6-7): 278-86.
         [http://dx.doi.org/10.1002/adma.19900020603]

[8]   Mishra R, Bhaskaran N, Ramkrishnan PA, Rao KJ. Lithium ion conduction in extreme polymer in salt
         regime. Solid State Ion 1998; 112(3-4): 261-73.
         [http://dx.doi.org/10.1016/S0167-2738(98)00209-4]

[9]   Baril D, Michot C, Armand M. The amoprphous nature of all polymer electrolyte films. Solid State
         Ion 1997; 94: 35.
         [http://dx.doi.org/10.1016/S0167-2738(96)00614-5]

[10]   Yu Q, Wang Y, Yi Z, *et al.* High Effieciency dye sensitized solar cells. ACS Nano 2010; 4: 6032.
         [http://dx.doi.org/10.1021/nn101384e] [PMID: 20923204]

[11]   Hauch A, Georg A. Diffusion in the electrolyte and charge-transfer reaction at the platinum electrode
         in dye-sensitized solar cells. Electrochim Acta 2001; 46(22): 3457-66.
         [http://dx.doi.org/10.1016/S0013-4686(01)00540-0]

[12]   Hao S, Wu J, Lin J, Huang Y. Modification of photocathode of dye-sensitized nanocrystalline solar
         cell with platinum by vacuum coating, thermal decomposition and electroplating. Compos Interfaces
         2006; 13(8-9): 899-909.
         [http://dx.doi.org/10.1163/156855406779366732]

[13]   Agrawal RC, Pandey GP. Solid polymer electrolytes: materials designing and all-solid-state battery
         applications: an overview. J Phys D Appl Phys 2008; 41(22): 223001.
         [http://dx.doi.org/10.1088/0022-3727/41/22/223001]

[14]   Han H, Bach U, Cheng YB, Caruso RA. Increased nanopore filling: Effect on monolithic all-soli-
         -state dye-sensitized solar cells. Appl Phys Lett 2007; 90(21): 213510.
         [http://dx.doi.org/10.1063/1.2743381]

[15]   Pandey K, Dwivedi MM, Singh M, Agrawal SL. Studies of dielectric relaxation and a.c. conductivity
         in [(100−x)PEO + xNH4SCN]: Al-Zn ferrite nano composite polymer electrolyte. J Polym Res 2010;
         17(1): 127-33.
         [http://dx.doi.org/10.1007/s10965-009-9298-3]

[16]   Zhang J, Han H, Wu S, *et al.* Conductive carbon nanoparticles hybrid PEO/P(VDF-HFP)/SiO$_2$
         nanocomposite polymer electrolyte type dye sensitized solar cells. Solid State Ion 2007; 178(29-30):
         1595-601.
         [http://dx.doi.org/10.1016/j.ssi.2007.10.009]

[17]   Nogueira AF, Durrant JR, De Paoli MA. Dye-Sensitized Nanocrystalline Solar Cells Employing a
         Polymer Electrolyte. Adv Mater 2001; 13(11): 826-30.
         [http://dx.doi.org/10.1002/1521-4095(200106)13:11<826::AID-ADMA826>3.0.CO;2-L]

[18]   Nogueira AF, Alonso-Vante N, De Paoli MA. Solid-state photoelectrochemical device using poly(o-
         methoxy aniline) as sensitizer and an ionic conductive elastomer as electrolyte. Synth Met 1999;
         105(1): 23-7.
         [http://dx.doi.org/10.1016/S0379-6779(99)00078-8]

[19]   Ileperuma OA, Dissanayake MAKL, Somasunderam S, Bandara LRAK. Photoelectrochemical solar
         cells with polyacrylonitrile-based and polyethylene oxide-based polymer electrolytes. Sol Energy
         Mater Sol Cells 2004; 84(1-4): 117-24.
         [http://dx.doi.org/10.1016/j.solmat.2004.02.040]

[20]   Freitas JN, Gonçalves AS, De Paoli M-A, Durrant JR, Nogueira AF. The role of gel electrolyte
         composition in the kinetics and performance of dye-sensitized solar cells. Electrochim Acta 2008;
         53(24): 7166-72.
         [http://dx.doi.org/10.1016/j.electacta.2008.05.009]

[21]   Flores IC, de Freitas JN, Longo C, De Paoli MA, Winnischofer H, Nogueira AF. Dye-sensitized solar
         cells based on TiO2 nanotubes and a solid-state electrolyte. J Photochem Photobiol Chem 2007;

189(2-3): 153-60.
[http://dx.doi.org/10.1016/j.jphotochem.2007.01.023]

[22]   Prabakaran K, Mohanty S, Nayak SK. Improved electrochemical and photovoltaic performance of dye sensitized solar cells based on PEO/PVDF–HFP/silane modified TiO_2 electrolytes and MWCNT/Nafion ® counter electrode. RSC Advances 2015; 5(51): 40491-504.
[http://dx.doi.org/10.1039/C5RA01770J]

[23]   Rao BN, Suvarna RP, Giribabu L, Raghavender M, Kumar VR. PEO based polymer composite with added acetamide, NaI/I2 as gel polymer electrolyte for dye sensitized solar cell applications. IOP Conf Series: Mate Sci Enger 2018; 310: 012012.

[24]   Joge P, Kanchan DK, Sharma P, Gondaliya N. Effect of nano filler on electrical properties of PVA-PEO blend polymer electrolyte. Indian J Pure Appl Phy 2013; 51: 350-3.

[25]   Tiautit N, Puratane C, Panpinit S, Saengsuwan S. Effect of SiO_2 and TiO_2 Nanoparticles on the Performance of Dye- Sensitized Solar Cells Using PVDF-HFP/PVA Gel Electrolytes. Energy Procedia 2014; 56: 378-85.
[http://dx.doi.org/10.1016/j.egypro.2014.07.170]

[26]   Venkatesan S, Liu IP, Chen LT, Hou YC, Li CW, Lee YL. Effects of TiO_2 and TiC Nanofillers on the Performance of Dye Sensitized Solar Cells Based on the Polymer Gel Electrolyte of a Cobalt Redox System. ACS Appl Mater Interfaces 2016; 8(37): 24559-66.
[http://dx.doi.org/10.1021/acsami.6b06429] [PMID: 27563731]

[27]   Limpan N, Prodpran T, Benjakul S, Prasarpran S. Influences of degree of hydrolysis and molecular weight of poly(vinyl alcohol) (PVA) on properties of fish myofibrillar protein/PVA blend films. Food Hydrocoll 2012; 29(1): 226-33.
[http://dx.doi.org/10.1016/j.foodhyd.2012.03.007]

[28]   Qiu K, Netravali AN. A Composting Study of Membrane-Like Polyvinyl Alcohol Based Resins and Nanocomposites. J Polym Environ 2013; 21(3): 658-74.
[http://dx.doi.org/10.1007/s10924-013-0584-0]

[29]   Aziz MF, Buraidah MH, Careem MA, Arof AK. PVA based gel polymer electrolytes with mixed iodide salts (K^+I^- and $Bu_4N^+I^-$) for dye-sensitized solar cell application. Electrochim Acta 2015; 182: 217-23.
[http://dx.doi.org/10.1016/j.electacta.2015.09.035]

[30]   Senthil RA, Theerthagiri J, Madhavan J. Hematite Fe_2O_3. Nanoparticles Incorporated Polyvinyl Alcohol Based Polymer Electrolytes for Dye-Sensitized Solar Cells. Mater Sci Forum 2015; 832: 72-83.
[http://dx.doi.org/10.4028/www.scientific.net/MSF.832.72]

[31]   Hassan HC, Abidin ZHZ, Chowdhury FI, Arof AK. High Efficiency Chlorophyll Sensitized Solar Cell with Quasi Solid PVA Based Electrolyte. Article, ID: Int. J. Photoene 2016; p. 3685210.

[32]   Seni RS, Puspitasari N. and Endarko (2017). Effect of the Addition of PEG and PVA Polymer for Gel Electrolytes in Dye-Sensitized Solar Cell (DSSC) with Chlorophyll as Dye Sensitizer. IOP Conf. Series: Materials Science and Engineering 214 (2017) 012011.

[33]   Chawla P, Tripathi M. Nanocomposite Polymer Electrolyte for Enhancement in Stability of Betacyanin Dye Sensitized Solar Cells. ECS Solid State Letters 2015; 4(6): Q21-3.
[http://dx.doi.org/10.1149/2.0041506ssl]

[34]   Katsaros G, Stergiopoulos T, Arabatzis IM, Papadokostaki KG, Falaras P. A solvent-free composite polymer/inorganic oxide electrolyte for high efficiency solid-state dye-sensitized solar cells. J Photochem Photobiol Chem 2002; 149(1-3): 191-8.
[http://dx.doi.org/10.1016/S1010-6030(02)00027-8]

[35]   Buraidah MH, Arof AK. Characterization of chitosan/PVA blended electrolyte doped with NH_4I. J Non-Cryst Solids 2011; 357(16-17): 3261-6.

[http://dx.doi.org/10.1016/j.jnoncrysol.2011.05.021]

[36]   Senthil RA, Theerthagiri J, Madhavan J, Arof AK. Performance characteristics of guanine incorporated PVDF-HFP/PEO polymer blend electrolytes with binary iodide salts for dye-sensitized solar cells. Opt Mater 2016; 58: 357-64.
[http://dx.doi.org/10.1016/j.optmat.2016.06.007]

[37]   Yousefi AA. Influence of Polymer Blending on Crystalline Structure of Polyvinylidene Fluoride. Iran Polym J 2011; 20: 109.

[38]   Wang R, Ahmed B, Raghuvanshi SK. Siddhartha, Sharma, N.P., Krishna, J.B.M. and Wahab, M.A. (2013). 1.25mev Gamma Irradiated Induced Physical and Chemical Changesin Poly Vinylidene Fluoride (PVDF) Polymer. Progress in Nanotechnology and Nanomaterials, 2, 42.

[39]   Noor MM, Buraidah MH, Yusuf SNF, Careem MA, Majid SR, Arof AF. Performance of Dye-Sensitized Solar Cells with (PVDF-HFP)-KI-EC-PC Electrolyte and Different Dye Materials. Article, ID: Int. J. Photoene 2011; p. 960487.

[40]   Perera KS, Vidanapathirana KP. Polyvinylidene fluoride based gel polymer electrolyte to be used in solar energy to electrical energy conversion. Sri Lankan Journal of Physics 2017; 17(0): 29-39.
[http://dx.doi.org/10.4038/sljp.v17i0.8037]

[41]   Kannadhasan S, Pandian MS, Ramasamy P. Synthesis of pure and 4-Nitroaniline doped (PVDF-HFP/LiI/I2) polymer electrolyte for dye sensitized solar cell (DSSC) applications. AIP Conf Proc 2017; 1832: 050061.
[http://dx.doi.org/10.1063/1.4980294]

[42]   Sun KC, Arbab AA, Sahito IA, *et al.* A PVdF-based electrolyte membrane for a carbon counter electrode in dye-sensitized solar cells. RSC Advances 2017; 7(34): 20908-18.
[http://dx.doi.org/10.1039/C7RA00005G]

[43]   Alves R, Donoso JP, Magon CJ, Silva IDA, Pawlicka A, Silva MM. Solid polymer electrolytes based on chitosan and europium triflate. J Non-Cryst Solids 2015; 432: 1.

[44]   Yusuf SNF, Azzahari AD, Yahya R, Majid SR, Careem MA, Arof AK. From crab shell to solar cell: a gel polymer electrolyte based on N-phthaloylchitosan and its application in dye-sensitized solar cells. RSC Advances 2016; 6(33): 27714-24.
[http://dx.doi.org/10.1039/C6RA04188D]

[45]   Aziz NA, Majid SR, Arof AK. Synthesis and characterizations of phthaloyl chitosan-based polymer electrolytes. J Non-Cryst Solids 2012; 358(12-13): 1581-90.
[http://dx.doi.org/10.1016/j.jnoncrysol.2012.04.019]

[46]   Muthumeenal A, Neelakandan S, Kanagaraj P, Nagendran A. Synthesis and properties of novel proton exchange membranes based on sulfonated polyethersulfone and N-phthaloyl chitosan blends for DMFC applications. Renew Energy 2016; 86: 922-9.
[http://dx.doi.org/10.1016/j.renene.2015.09.018]

[47]   Buraidah MH, Teo LP, Yusuf SNF, *et al.* TiO_2 / chitosan-NH_4I(+I2)-BMII-based dye-sensitized solar cells with anthocyanin dyes extracted from black rice and red cabbage. Int J Photoenergy 2011; 2011: 1-11.
[http://dx.doi.org/10.1155/2011/273683]

[48]   Yusuf SNF, Aziz MF, Hassan HC, *et al.* Phthaloylchitosan-based gel polymer electrolytes for efficient dye-sensitized solar cells. J Chem 2014; 2014: 1-8.
[http://dx.doi.org/10.1155/2014/783023]

[49]   Maddu A, Gusra H, Indro MN. Development of solid-state photoelectrochemical solar cell circuits based on chitosan-PEG blend electrolyte. Int J Ren Ene Res 2017; 101: 7.

[50]   Chen C, Liao JY, Chi Z, et al. Metal-free organic dyes derived from triphenylethylenefor dye-sensitized solar cells: tuning of the performance by phenothiazine and carbazole. J Mat Chem 2012; 22: 8994-9005.

SUBJECT INDEX

A

Absorption 8, 28, 31, 76, 135
  infrared 31
   of photoenergy 8
  sound 76
Acid 70, 72, 73, 83, 87, 112, 121, 123, 128
  adipic 83, 87
  amino 128
  citric 112
  diamine 83
  nitric 70
  nucleic 121, 123
  sulphuric 70
  terephthalic 72, 73
Agrobacterium rhizogenes 128
Applications 36, 41, 67, 77, 79, 80, 81, 82, 90, 96
  automotive 81
  biomedical 77, 96
  industrial 41, 81, 82, 90
  medical 67, 79, 80
  of chalcogenide glasses 36
Atomic 28, 46, 95, 97, 141, 144, 148, 151, 152, 166
  force microscopy (AFM) 28, 95, 97, 166
  layer deposition (ALD) 46, 141, 144, 148, 151, 152

B

Biosensors 96, 150
  nano-tubes-based 150
Brownian 100, 101, 102
  motion 100, 102
  particle 101

C

Capacitors, electrochemical 144, 151
Carbon 48, 125, 141, 148

nanoparticles 141
nanotubes 48, 125, 148
Cardiovascular disease 127
Cation binding affinities 126
Cellulose 63, 72, 121
  dissolving wood 72
  nanomaterials 121
  regenerated 63
Cellulose acetate 63, 65, 69
  fibre 69
Chalcogenide materials 36, 38
Chemical 15, 18, 19, 51, 121, 141, 148
  bath deposition (CBD) 15, 19
  separation methods 121
  vapour deposition (CVD) 15, 18, 19, 51, 141, 148
Chitosan 168, 169
  based polymer electrolyte 168
  blending 169
Clothing 66, 67, 71, 78, 80, 85, 87, 88, 92
  flame-resistant 87
Coatings, welding rod 96
Cohesive forces 35
Conducting electrodes, transparent 49, 52
Conduction, low electrical 39
Conductors, transparent 52, 53
Contaminants 87, 123
  toxic 123
Contaminated water 120, 126
Contamination 121
  radioactive 121

D

Detrended fluctuation analysis (DFA) 97
Devices 4, 6, 36, 40, 51, 55, 56, 57, 78, 96, 111, 145, 146, 153, 158, 159, 160
  commercial magnetic memory 111
  electrochemical 159, 160
  energy storage and conversion 145, 153
  medical 78
  memory 4

www.ingramcontent.com/pod-product-compliance
Lightning Source LLC
Chambersburg PA
CBHW041702210326
41598CB00007B/503